ND DARWIN FAMILIES

Wedgwood
4–1815

Sarah Elizabeth 1778–1856

Josiah II (Uncle Jos) 1769–1843 = Elizabeth (Bessy) Allen 1764–1846

- Sarah Elizabeth (Elizabeth, Aunt Sarah) 1793–1880
- Mary Anne 1796–8
- Charlotte 1797–1862 = Charles Langton 1801–86
  - Edmund 1841–75
- Henry Allen (Harry) 1799–1885 = Jessie Wedgwood 1804–72
  - Louisa Frances 1834–1903
  - Caroline b. 1836
  - John Darwin 1840–70
  - Anne Jane 1841–77
  - Arthur 1843–1900
  - Rowland 1847–1921
- Francis 1800–88 = Frances Mosley 1808–74
  - Godfrey 1833–1905
  - Amy 1835–1910
  - Cicely Mary 1837–1917
  - Clement 1840–89
  - Lawrence 1844–1913
  - Constance Rose 1846–1903
  - Mabel Frances 1852–1930
- Hensleigh 1803–91 = Frances (Fanny) Mackintosh 1800–89
  - Frances Julia (Snow) 1833–1913
  - James Mackintosh (Bro) 1834–64
  - Ernest Hensleigh 1838–98
  - Katherine Euphemia (Effie) 1839–1931
  - Alfred Allen 1842–92
  - Hope Elizabeth 1844–1934
- Frances (Fanny) 1806–32

5
2
1927
–1926

Support for the editing of volumes 7 to 16 of the *Correspondence of Charles Darwin*, from which letters from 1860 to 1868 have been extracted, and for the editing of this volume of the *Selected letters of Charles Darwin* has been received from the National Endowment for the Humanities, the National Science Foundation, the Andrew W. Mellon Foundation, the Alfred P. Sloan Foundation, the Royal Society of London, the British Academy, the Pilgrim Trust, the Isaac Newton Trust, the Wellcome Trust, the Pew Charitable Trusts, the Natural Environment Research Council, the British Ecological Society, the Stifterverband für die Deutsche Wissenschaft.

The National Endowment's grants (Nos. RE-23166-75-513, RE-27067-77-1359, RE-00082-80-1628, RE-20166-82, RE-20480-85, RE-20764-89, RE-20913-91, RE-21097-93, RZ-20393-99, RZ-20849-02, RE-21282-95, RZ-20018-97, and RQ-50154-05) were from its Program for Editions; the National Science Foundation's funding of the work was under grants Nos. SOC-75-15840, SES-7912492, SES-8517189, SBR-9020874, SES-0135528, SOC-76-82775, SBR-9616619, and SES-0646230.

Charles Darwin
Photograph by William Erasmus Darwin, 1864.
By permission of the Syndics
of Cambridge University Library.

Evolution

SELECTED LETTERS OF CHARLES DARWIN 1860–1870

EDITED BY

FREDERICK BURKHARDT
SAMANTHA EVANS
ALISON M. PEARN

FOREWORD BY DAVID ATTENBOROUGH

CAMBRIDGE UNIVERSITY PRESS
Cambridge, New York, Melbourne, Madrid, Cape Town, Singapore, São Paulo, Delhi

Cambridge University Press
The Edinburgh Building, Cambridge CB2 8RU, UK

Published in the United States of America by Cambridge University Press, New York

www.cambridge.org
Information on this title: www.cambridge.org/9780521874120

First published 2008

Printed in the United Kingdom at the University Press, Cambridge

A catalogue record for this publication is available from the British Library

Library of Congress Cataloguing in Publication data

ISBN 978-0-521-87412-0 hardback

Dedicated to
the Wellcome Trust
for their generous support of
the Darwin Correspondence Project
over many years

Contents

Foreword

SIR DAVID ATTENBOROUGH

OM CH FRS

The letters that Charles Darwin wrote immediately after the publication of *On the Origin of Species* are not as well known as his journal written during the voyage of the *Beagle* or those he wrote during the long period between his return home and the book's appearance. But they deserve to be, for they shed important light on many fascinating questions. How did he react to the sensation that his revolutionary book created? What were his religious views that he so carefully refrained from explaining in public? And what kind of life did he lead, shut away in his house in the depths of Kent, apparently cut off from his scientific colleagues in London and elsewhere?

But first, why did he not attend any of the scientific debates in London in which his thoughts on evolution played such an important part? These letters certainly make the answer to that question very clear. He was, truly, a chronically sick man. The nature of his illness may or may not have had a psychosomatic component but of its reality and severity there can be no doubt. In letter after letter he refers to his daily sicknesses. A journey will cause severe vomiting; the prospect of having to speak publicly will leave him prostrate.

But cut off though he was physically, the postal service kept him in touch with a wide range of correspondents. By this time, the penny post was well established and no doubt the postman in Downe village called at Down House several times a day. He brought letters from far afield, from Kew and the Isle of Wight, Edinburgh and Aberdeen, Magdeburg and Geneva, Massachusetts and Illinois, even West Africa and Indonesia. Charmingly, Darwin asks Asa Gray, the Harvard botanist, to send him some

American stamps and specifies exactly which ones he wants. As a boy, Darwin had obsessively collected beetles. Now, his own son has become a stamp-collector. And surprisingly perhaps, since it is not exactly in tune with the image of staid pillars of the Victorian scientific establishment, he and his correspondents send their portrait photographs to one another.

Darwin himself writes as many as ten letters a day. They are full of questions. Are blue-eyed rabbits deaf? Does a capuchin monkey open its mouth when surprised? Does the colour of the beards of Slavonic men differ from that of their hair? The minuteness of the detail he demands should not surprise us. Detail, accumulated, marshalled, classified and distilled, is of course the very foundation of natural science. More unexpected is the extraordinarily wide range of his curiosity. He ponders on the glacial origins of the mysterious parallel roads of Glen Roy; he speculates about the mathematics that explain the angles at which leaves in a spire sprout from a stem, and learns about the antiquity of Stone Age hand axes.

On one subject, about which we might have expected a great deal of comment, he remains almost silent. The year after the publication of *Origin*, the first fossil skeleton of *Archaeopteryx* was discovered in Bavaria. It combined the characteristics of reptiles and birds and was purchased by the Natural History Museum in London. A supporter of Darwin's, the palaeontologist Hugh Falconer, writes excitedly to Darwin about it, for this is exactly the link between great groups of the animal kingdom that readers of *Origin* might have hoped to find in support of Darwin's theory. But Darwin himself makes little comment. Richard Owen, the Director of the Museum who first described the fossil, is a virulent anti-evolutionist and harbours a deep personal enmity for Darwin. Perhaps it is that that prevents the gentle pacific Darwin from indulging in any triumphalism.

His correspondence with Alfred Russel Wallace is particularly interesting. The two men had put their names jointly to the first announcement of the theory of natural selection at a meeting of the Linnean Society. Darwin had been collecting evidence and thinking about the problem for a quarter of a century, Wallace for a matter of months. The acceptance of each by the other as a joint author is one of the most heart-warming examples of generosity of spirit in the annals of scientific discovery. But was it achieved with a mental grinding of teeth? Were both men as genuinely generous privately as their public acts suggest. Their letters, included here, make it clear that indeed they were.

The two could hardly have come from more different backgrounds. Wallace had had to pay for his travels by collecting entomological specimens for sale. Darwin was a wealthy member of the landed gentry. But Wallace had no hesitation in acknowledging Darwin's precedence.'Do not call the theory mine,' he writes to Darwin, 'It is truly yours.' But that does not prevent him from criticising Darwin for adopting the phrase 'natural selection'. 'Survival of the fittest', he argues, would he more accurate. And Darwin, for his part, when one of his other correspondents points out to him that Wallace is beginning to 'backslide' when it comes to applying natural selection to humanity, remains publicly silent.

Of all his correspondents Darwin seems most at ease and most intimate with the botanist Joseph Hooker. He teases Hooker about his 'crockery madness'—for Hooker is an avid collector of Wedgwood china. Hooker gives Darwin a racy, gossipy account of the famous routing of Bishop Wilberforce at the meeting of the British Association for the Advancement of Science in Oxford. And Darwin feels uninhibited enough to boast to Hooker about the sales of his books. The two men ridicule to one another the absurdity of the Duke of Argyll's explanation of the survival of rudimentary organs, and Hooker explains that he could not go into the relationship of science and religion when he dines with the Duke—at least not in front of the Duke's wife and children.

As to Darwin's religious convictions and doubts, there are no direct statements but several indirect inferences. 'I cannot persuade myself,' he writes to Asa Gray, 'that a beneficent & omnipotent God would have designedly created the Ichneumonidæ with the express intention of their feeding within the living bodies of caterpillars'.

The last of the letters in this volume were written in 1870. Darwin still had twelve years of observation and pondering ahead of him. *Origin* was not, as might be supposed, his final work, the summation of a life-time's thought. On the contrary. it was only the second of his scientific books written for the general public, having been preceded by his account of his travels in the *Beagle* and his study of coral reefs. Eight major works were yet to come. The letters printed here give clues to what these will be. He writes to correspondents overseas to enquire about facial expressions of different human races. In his greenhouse, he analyses the climbing techniques of plants and the fertilisation mechanisms of orchids, and watches sundew under his lens catching insects on its sticky hairs. He sends questionnaires to breeders of pigeons and fancy fowl. Each of these subjects, in due course, would be examined in magisterial detail in one of the fat green volumes published by Mr Murray in London.

But Darwin is also aware that his theory of evolution by natural selection lacks one key element. By what mechanism are the characteristics of an individual transferred to the next generation? Darwin outlines to several of his correspondents a theory he calls 'pangenesis'. This suggests that individual organs can secrete minute particles he calls 'gemmules' into the blood-stream which can enable some animals to regenerate their limbs. His cousin, Francis Galton, tests the theory for him by injecting blood from rabbits with white markings into a pregnant silver-grey doe, to see if hair colour can be transferred in this way. But convincing proof is lacking and Darwin goes on worrying about it. He reverts to it again and again, but was never to find the solution to this problem. At precisely the time that Darwin was writing these letters, an obscure monk in central Europe was investigating heredity by breeding generation after generation of garden peas in his monastery garden. Would these observations have helped or hindered Darwin in his own speculations about the mechanisms of inheritance? It certainly took half a century of work by twentieth-century biologists to reconcile the two men's discoveries. How one wishes that Mendel had made as much use of the postal service as Darwin did.

Introduction

The decade immediately following publication of *On the origin of species* in 1859 to the eve of publication of *Descent of man* in 1871 was arguably the most intense and productive of Charles Darwin's life. These were years in which the implications of the theories made public through *Origin* were explored and debated around the world, not only in the scientific community but in the public arena. Darwin, so far as his health would allow, set about countering criticisms with ever more detailed researches into complex mechanisms in organisms, teasing out how they could be explained as adaptations arising through the operation of natural selection. He also sought answers to the questions he knew *Origin* had not answered, in particular concerning the mechanisms of inheritance, and the evolutionary role of competition for sexual partners.

At the beginning of this period Darwin still intended to write the 'Big Book' on species of which *Origin* was only an abstract. As he resumed work on what had been intended as a single chapter on pigeon-breeding, however, it quickly became apparent that a detailed exposition of the production of domestic varieties of the various animals he was researching would require a separate publication. In fact as his researches deepened and widened publications expanded out of one another like Russian dolls: a planned final chapter on human origins for *Variation under domestication* became another two-volume work, *Descent of man and selection in relation to sex*, and his work on the relationship of human and animal emotions outgrew the confines of *Descent* and was eventually published in 1872 as *Expression of the emotions in man and animals.*

The year 1859 had ended well for Darwin, who had been delighted at the news that the first edition of *Origin* had sold out on its first day, and was both relieved and deeply gratified by the generally positive initial reaction of his scientific colleagues. It was the more critical responses that were to set the tone for the next decade, however. Darwin's major task in these years was to respond in detail, privately through letters and publicly through a rapid series of new editions and further publications, to both the specific scientific arguments raised against his theory and the philosophical qualms it inspired. A second, revised, edition of *Origin* was already out in January 1860, and the third substantially updated edition by the end of the year. By 1870, Darwin had published a fifth edition and was already gathering material for the sixth and final edition, published in 1872.

There was support for Darwin's theories in a series of significant publications by others. To Darwin's satisfaction, Henry Walter Bates's work on protective mimicry invoked the mechanism of natural selection to explain the development of complex markings in insects. Thomas Henry Huxley's *Evidence as to man's place in nature* was published in 1863, as was Charles Lyell's *Antiquity of man*, although this disappointed Darwin in its cautious stance.

Darwin keenly followed the debates about his work. His ideas gained ground throughout the 1860s, as the numerous honorary fellowships and degrees bestowed upon him attest, but their acceptance was not achieved without struggle and was not wholesale. The annual award of the Royal Society of London's Copley Medal became a battleground between Darwin's supporters, such as Thomas Huxley and Hugh Falconer, and those members of the scientific establishment who, while respecting many of Darwin's achievements, were anxious that the society not be seen to endorse the theories contained in *Origin*. Unsuccessful nominations in 1862 and 1863 were followed by a narrow and controversial victory in 1864.

Critical reviews of *Origin* appeared abroad also, in particular in France, and many naturalists, such as the botanist Charles Naudin, although continuing to assist Darwin in his research, remained unconvinced by his arguments. In Germany, Darwin's theories were spread by younger scholars such as Ernst Haeckel but resisted, to Darwin's distress, by others. Darwin's revisions to the fifth edition of *Origin* were made largely in direct response to criticisms such as those from the botanist Carl von Nägeli.

One criticism that Darwin was very conscious could be levelled at the arguments in *Origin* was the absence of any explanation of how inheritance worked. He countered this with his theory of 'pangenesis', first privately circulated in 1865 but not published until 1868, when it appeared in *Variation.* Darwin postulated that 'gemmules' present in bodily fluids and transmitted from parent to child had the ability to develop into different parts of organisms, but could lie dormant from generation to generation. Painfully conscious of the difficulty of supporting this theory with evidence, Darwin was disappointed, but not surprised, by its mixed reception.

Darwin continued to be deeply interested in all questions concerning heredity, encouraging his cousin, Francis Galton, in his experiments transfusing blood in rabbits, and even suggesting that a question on cousin-marriage be inserted in the national census to gather data on the effects of inbreeding in humans.

Darwin had more success with the development of his theories concerning sexual selection. One of the arguments raised against natural selection was its inability to account for beauty in nature where that beauty apparently failed to offer any survival advantage. Much of Darwin's correspondence in this period reflects his interest in gathering evidence of the importance of colour, sound, and smell in the attraction of sexual partners. He debated the relative importance of the mechanisms of sexual selection and natural selection in correspondence with Alfred Russel Wallace, who was inclined to attach less significance to the operation of sexual selection than was Darwin.

Objections to Darwin's theory on religious grounds, and the question of the origin of humankind, are tackled directly in the correspondence from the beginning of this period in a way that Darwin did not attempt in public until the publication of *Descent of man* more than ten years later. Darwin and a select few of his correspondents, including Asa Gray and Alfred Russel Wallace, privately debated the role natural selection and sexual selection might have played in human development, against a background of public tension over the impact of new scientific discoveries and new critical methods on traditional religious belief. That tension had been heightened by the publication in 1860 of *Essays and reviews*, a collection of essays that addressed the implications of recent scholarship for traditional religious interpretations, and was severely attacked by the Church establishment; it was further sustained in 1863 by the trial for heresy of John William Colenso, the liberal bishop of Natal, to whose defence fund Darwin contributed. Colenso advocated a commonsense, critical approach to the study of the Bible; his examination of the Old Testament had caused great controversy because of its denial that Moses was the author of the Pentateuch.

Although careful to avoid open confrontation himself, Darwin in these years promoted works he thought would support his case. In 1860 he arranged the re-publication in England of reviews of *Origin* by the Harvard botanist and devout Presbyterian Asa Gray, who argued that the theory of natural selection was not incompatible with belief in design in nature. Darwin also ensured that translations of his own works appeared quickly and fostered links with supporters such as Julius von Haast in New Zealand and Benjamin Dann Walsh in the United States, and, most notably, with Fritz Müller, a German naturalist and teacher living in Brazil, whose book *Für Darwin*, a study of the developmental history of the Crustacea, which Müller presented as a validation of Darwin's theory, was translated into English in 1869 at Darwin's own expense.

Because of the increasing importance to Darwin's work in these years of communications received through correspondence, the editors have included both letters he received and those written by him in this volume. Darwin's network of correspondents expanded rapidly after the publication of *Origin.* This expansion was a result of Darwin's increased fame, but also of his conscious use of letter-writing to elicit the mass of observations about the natural world that he needed to underpin his arguments. Darwin's study of human emotion led him to devise a questionnaire that spread out like a chain letter from correspondent to correspondent to reach remote outposts of the British Empire and beyond. Correspondents such as the pigeon-breeder William Bernhard Tegetmeier supplied Darwin with specimens and drawings, as well as a great deal of information that was incorporated into *Variation.* Darwin even used correspondence to coordinate experiments undertaken around the world on his behalf. John Scott, a gardener at the Royal Botanic Garden in Edinburgh, and later in Calcutta, undertook experiments on sterility between plant varieties at Darwin's instigation. Fritz Müller in Brazil, and his brother Hermann in Germany, replicated botanical experiments that Darwin was undertaking at Down. In return Darwin provided encouragement and patronage to this informal research circle—he helped many, such as Scott and George Henslow, to publish in their own right, and he used his influence to promote the careers of Scott and William Sweetland Dallas, the indexer of *Variation.*

In his own observational and experimental work, Darwin concentrated increasingly on botanical subjects. He realised that the results of experiments investigating cross-breeding could be achieved much more quickly in plants than in animals, and, although careful never to claim expertise in botany, he published a series of pioneering papers.

In 1860, Darwin was already studying adaptive mechanisms in carnivorous plants and explaining their development by reference to natural selection, and his letters reveal that he had also begun to appreciate the importance of intercrossing between individuals within a species through his work on *Primula*. He published on fertilisation mechanisms in orchids in 1862, and wrote up his conclusions about plants in which the flowers take different forms in a series of papers throughout the 1860s. 'Dimorphic condition in *Primula*' was published in 1862, and 'Three forms of *Lythrum salicaria*' in 1864. A hothouse that Darwin had built in his garden in 1863 allowed him to extend his observations and experiments across a wider range of plants, many provided by his friend Joseph Hooker, director of the Royal Botanic Garden at Kew, and gave him many happy hours among his beloved orchids and carnivorous plants. The hothouse also allowed him to follow a suggestion from Asa Gray and begin a study of the mechanisms by which plants twine and climb. Darwin's explanation of the development of these mechanisms in many families of plants was published only two years later in 1865 in *Climbing plants.*

Many of these botanical researches began as pastimes, in particular while Darwin was ill and unable to do other work. By the middle of 1863 Darwin was in increasingly poor health. Advised by a succession of doctors, he tried a series of treatments and regimes, but only finally recovered to some extent in 1866. Ever alert to opportunities, Darwin used many of the same doctors to provide him with scientific data, especially as he turned his attention finally and fully to the study of human development that led to the publication of *Descent* and *Expression of the emotions.*

The decade covered by this volume saw momentous political events, including both the beginning and the end of the American Civil War. Following the abduction by Union soldiers early in 1861 of Confederate envoys travelling on board a British mail packet, it appeared to Darwin's distress that Britain might declare war on the Union. He commented on the course of the war in his letters to Asa Gray, and, as a passionate opponent of slavery, shared Asa Gray's jubilation at the Union victory of 1865.

At home, Darwin watched his seven surviving children grow to adulthood. In 1860, the youngest, Horace, was 9 years old, and the eldest, William, was 21, and casting about for a career. By 1870, William was an established partner in a bank in Southampton, his brother George, to his parents' delight, had graduated from Cambridge University with the second highest place in the final mathematics examinations, and their third son, Francis, was a student and falling worryingly into debt. It was against this domestic background that Darwin carried out his scientific work, and his children continued to contribute to it at home and elsewhere: William sent observations on plants from the south coast, George helped with mathematical calculations, and Darwin's daughter Henrietta, one of the few people Darwin entrusted with reviewing the manuscript of *Descent*, emerges from Darwin's correspondence at the end of this period as an unsung but important influence on his continuing researches.

Editors' note

The letters in this volume have been selected from Darwin's correspondence in the years between 1860 and 1870, that is, from just after the publication of *Origin of species* to just before the publication of *Descent of man*. At the time of writing, letters from 1860 to 1867 had been published in full in volumes 8 to 15 of the *Correspondence of Charles Darwin* (CUP, 1993–2005). As with the first volume of selected letters (*Charles Darwin's letters: a selection 1825–59*, CUP 1996, revised edition 2008), the editors have sought to make the selection representative of the larger work and to provide a trustworthy portrayal of Darwin's mind, personality, and method of work. This volume of selected letters includes letters to Darwin as well as letters from him, enabling readers to follow the developing dialogues between Darwin, his friends, family, and readers.

The texts of the selected letters for the most part retain the spelling and punctuation of the originals. Editorial interpolations and corrections, made only where the text would otherwise be hard to understand, are placed in square brackets, and all editorial omissions are indicated by ellipses. Where Darwin used Down headed notepaper to write from another address, his true location has been added to the address in square brackets. Square brackets have also been used to disambiguate addresses, for example 'Cambridge [Massachusetts]'.

Italics are used in the transcriptions to indicate printed letterhead in the originals and also to indicate underlining by the writer. Bold-face type is used to indicate multiple underlining by the writer. Headnotes and footnotes are set in smaller type than that used for the letter texts.

Letters never intended for publication are likely to contain references that require explanation. The notes to the letters have been adapted from the fully annotated *Correspondence*, where possible, with the interests of the general reader in mind. The letters are followed by a biographical list of the individuals mentioned in the letters, as well as a bibliographical note on Darwin's writings mentioned in this volume. Letters in languages other than English have been translated.

In the headings and notes throughout the volume, Darwin is referred to as CD.

The editors would like to express their thanks to Ellis Weinberger and Matt Daws for help with typesetting; to Nicholas Gill for help with proofreading previously unpublished texts; to Margot Levy for the index; and to Rosemary Clarkson for research assistance. The initial selection of letters was carried out with the help of Ruth Goldstone. Where letters have been previously published in volumes of the *Correspondence of Charles Darwin*, the notes added here obviously rely heavily on the work of the editors of those volumes, whom we also wish to thank.

Frederick Henry Burkhardt (1912–2007)

It is with great sadness that we record the death on 23 September 2007 of Fred Burkhardt, the founding editor of the Darwin Correspondence Project. Fred died shortly before this volume went to press. He remained actively engaged in the Project up to the last weeks of his life and will be deeply missed by all of us who have been privileged to count him as a colleague and a friend.

Symbols and abbreviations

[some text]	'some text' is an editorial insertion or correction
*[*some text*]*	'some text' is a conjectured reading of an ambiguous word or passage
⟨ ⟩	a destroyed word or words
⟨some text⟩	'some text' is a suggested reading for a destroyed word or passage
CD	Charles Darwin

Prologue

The first letter in this volume is a response to *Origin of species*, which was published in 1859. The following notes give some details of Darwin's life up to that point.

1809
Charles Darwin was born on 12 February, second son and fifth of six children of Robert Waring Darwin, a physician in Shrewsbury, and Susannah, daughter of Josiah Wedgwood I, master-potter of Staffordshire.

1818
Entered Shrewsbury School as a boarding student.

1825
Went to Edinburgh with his brother Erasmus to study medicine.

1828
Entered Christ's College, Cambridge, to study for a degree with the intention of becoming a clergyman.

1831
At the end of the year, sets sail in HMS *Beagle* as companion to Captain FitzRoy on a survey voyage to South America and the Pacific.

1836
Arrived back in England in October.

1838
Publication of the five-volume *Zoology of the Beagle* begins.

1839
Married his cousin, Emma Wedgwood. *Voyage of the Beagle* published, under the title *Journal and remarks*.

1842
Charles and Emma move with their two children from London to Down in Kent. Publication of *Coral reefs*.

1844
Publication of *Volcanic islands*. Darwin leaves instructions with Emma for the publication of a sketch of his 'Species theory' in the event of his untimely death.

1846
Publication of *Geological observations on South America*.

1851
Charles and Emma's eldest daughter, Anne, dies aged 10. Publication of Darwin's research on barnacles begins.

1858
Alfred Russel Wallace sends Darwin a manuscript describing his independent discovery of natural selection. Darwin's friends counsel the presentation of a joint paper at the Linnean Society.

1859
Publication of *On the origin of species by means of natural selection* on 24 November.

1860

[This is one of the earliest letters of criticism of the *Origin* that CD received. Leonard Jenyns was a long-time friend of CD's and had been an early rival in entomological collecting in CD's Cambridge student days. Jenyns had also identified the fish that CD had collected (see *Zoology of the voyage of HMS Beagle*, Part IV *Fish*, edited by CD).]

From Leonard Jenyns 4 January 1860[1]

Jan. 4. 1860.

My dear Darwin

I have read your interesting book with all carefulness as you enjoined,—have gleaned a great deal from it, & consider it one of the most valuable contributions to Nat. Hist Literature of the present day. . . .

But I frankly confess I did not look for any such large assemblages of species to be brought together in this way, as the descendants from one & the same stock, similar to what you have attempted in your volume. By this you will see that I embrace yr theory in part, but hardly to the full extent to which you carry it. Still I allow you have made out a very strong case, and I will not pretend to say what future researches in the same direction may not ultimately establish.

I can quite fall in with the view that those fossil animals which so closely resemble their living representatives at the present day, are in fact the progenitors of these last;—such indeed has been my opinion for many years, tho' a contrary one, I know had been adopted by many of our first Geologists & Naturalists. I can also well believe that whole families have had a common parentage at some remote period of the past,—& that the same *may* have been the case, reasoning analogically (tho' this is not always safe in Nat. Hist. speculations),—with groups of even higher denomination.

But I cannot think that *all* the difficulties which stand in the way of so extensive a generalization have been entirely got over—

It seems to me that if "all organic beings that have ever lived on the earth, had descended from some one primordial form", as you seem to think possible,—we should find (either among the fossil or living species) the same connecting links, & "fine gradations" between the highest groups, that we *do* find among the lower, & which renders classification & definition so difficult.[2] Thus birds as a class may be shown to be all more & less closely connected, with only here & there a saltus which we cannot hedge over: the same perhaps of all vertebrate animals—if we had all the lost & unknown living forms at hand to fill up the gaps;—but when we come to another quite distinct type of organization—the annulose, for instance, how little of intermediate between this & the vertebrate, or between either of these & the molluscous, &c. Still more what an hiatus between vegetables & animals, tho' just at their respective origins, or where the vital powers seems least active, they are scarcely separable by any distinguishing characters. But why should this never be the case afterwards—& not merely at the beginning of things—if they originally sprung from the same primordial germ?

One great difficulty to my mind in the way of your theory is the fact of the existence of Man. I was beginning to think you had entirely passed over this question, till almost in the last page I find you saying that 'light will be thrown on the origin of man & his history'.[3] By this I suppose is meant that he is to be considered a modified & no doubt *greatly* improved orang! I doubt if this will find acceptance with the generality of readers— I am not one of those in the habit of mixing up questions of science & scripture, but I can hardly see what sense or meaning is to be attached to *Gen*: 2.7. & yet more to vv. 21. 22, of the same chapter, giving an account of the creation of *wo* man,—if the human species at least has not been created independently of other animals, but merely come into the world by ordinary descent from previously existing races—whatever those races may be supposed to have been. Neither can I easily bring myself to the idea that man's reasoning faculties & above all his *moral sense,* c^{d}. ever have been obtained from irrational progenitors, by mere natural selection—acting however gradually & for whatever length of time that may be required. This seems to be doing away altogether with the

Divine Image which forms the insurmountable distinction between man & brutes.

[1] The original letter has not been found. CD apparently sent it to his friend, the geologist Charles Lyell, who copied it into his journal with the heading: 'Letter from Rev^d. Leonard Jenyns to C. Darwin, Jan. 4. 1860.'
[2] The quotations are from *Origin*, p. 484.
[3] *Origin*, p. 488.

To Leonard Jenyns 7 January [1860]

Down Bromley Kent
Jan. 7th

My dear Jenyns

I am very much obliged for your letter. It is of great use & interest to me to know what impression my Book produces on philosophical & instructed minds. I thank you for the kind things which you say; & you go with me much further than I expected.— You will think it presumptuous, but I am convinced, *if circumstances lead you to keep the subject in mind*, that you will go further. No one has yet cast doubt on my explanation of the subordination of group to group, on homologies, Embryology & Rudimentary organs; & if my explanation of these classes of facts be at all right, whole classes of organic beings must be included in one line of descent.— The imperfection of the geological Record is one of greatest difficulties (by the way, Lyell, who is convert, does not think that I have exaggerated imperfection). During earliest period the record would be most imperfect, & this seems to me sufficiently to account for our not finding intermediate forms between the classes in the same great Kingdoms [of organic life].—

It was certainly rash in me putting in my belief of probability of all beings having descended from *one* primordial form; but as this seems yet to me probable, I am not willing to strike it out.— Huxley alone supports me in this, & something could be said in its favour.—

With respect to man, I am very far from wishing to obtrude my belief; but I thought it dishonest to quite conceal my opinion.— Of course it is open to everyone to believe that man appeared by separate miracle, though I do not myself see the necessity or probability.—

Pray accept my sincere thanks for your kind note. Your going some way with me gives me great confidence that I am not very wrong. For a very long time I halted half-way; but I do not believe that any enquiring mind will rest half-way. People will have to reject all or admit all,—by *all* I mean only the members of each great Kingdom.—

My dear Jenyns | Yours most sincerely | C. Darwin

[Huxley, in a lecture at the Royal Institution of Great Britain, stated his general approval of CD's work, but nonetheless said that it fell short of being a satisfactory theory, primarily because there was as yet no proof that varieties that could not produce fertile offspring together (i.e., that were physiologically different species) had ever been produced from a common stock.]

To Thomas Henry Huxley 11 January [1860]

Down Bromley Kent
Jan. 11th

My dear Huxley

I fully agree that the difficulty is great, & might be made much of by a mere advocate. Will you oblige me by reading again slowly from [*Origin,*] p. 267–272.— I may add to what is there said, that it seems to me quite hopeless to attempt to explain why varieties are not sterile; until we know precise cause of sterility in species.— Reflect for a moment on how small & on what very peculiar causes the unequal *reciprocity* of fertility in same two species must depend.— Reflect on the curious case of species *more* fertile with foreign pollen than their own. Reflect on many cases which could be given, & shall be given in my larger book[1] (independently of hybridity) of very slight changes of conditions causing one species to be quite sterile & not affecting a closely allied species.— How profoundly ignorant we are on this intimate relation between conditions of life & impaired fertility in pure species.— ...

Reflect on case of the vars. of Verbascum which differ in no other respects whatever besides the fluctuating element of colour of flower, & yet it is impossible to resist Gärtner's evidence, that this difference in the colour does affect the mutual fertility of the varieties. The whole case seems to me far too mysterious to rest valid attack on the theory of modification of species, though, as you say, it offers excellent ground for mere advocate.—

I am surprised considering how ignorant we are on very many points, that more weak parts in my Book have not as yet been pointed out to me. No doubt many will be. . . .

My dear Huxley | Yours most sincerely | C. Darwin

[1] CD was at work on *Variation of animals and plants under domestication*, the first part of a planned larger work on natural selection.

From Asa Gray 23 January 1860

Cambridge [Massachusetts].
Jany. 23, 1860

My Dear Darwin . . .

Such little notices in the papers here as have yet appeared are quite handsome and considerate.

I hope next week to get printed sheets of my review from New Haven and send to you, and will ask you to pass them on to D^r. Hooker.[1]

To fulfil your request I ought to tell you what I "think the weakest, & what the best parts of your book. But this is not easy, nor to be done in a word or two. The *best part*, I think, is the *whole*, i.e. its *plan* & *treatment*,—the vast amount of facts and acute inferences handled as if you had a perfect mastery of them. I do not think 20 years too much time to produce such a book in.

Style clear & good, but now & then wants revision for little matters. (p. 97, self-fertilises *itself*—&c)

Then your candor is worth everything to your cause. It is refreshing to find a person with a new theory who frankly confesses that he finds difficulties—insurmountable, at least for the present. I know some people who never have any difficulties, to speak of.

The moment I understood your premises, I felt sure you had a real foundation, to build on. Well, if one admits your premises, I do not see how he is to stop short of your conclusions, as a probable hypothesis, at least. It naturally happens that my review of your book does not exhibit any thing like the full force of the impression the book has made upon me. Under the circumstances I suppose I do your theory more good here, by bespeaking for it a fair and favorable consideration, and by standing non-committal as to its full conclusions, than I should if I announced myself a convert,—nor could I say the latter, with truth.

Well, what seems to me the weakest point in the book is the attempt to account for the formation of organs,—the making of eyes, &c by natural selection. Some of this reads quite Lamarckian.

The Chapter on *Hybridism* is not a *weak*, but a *strong* chapter. You have done wonders, there. But still you have not accounted as you may be held to account, for divergence up to a certain extent producing increased fertility of the crosses—but carried one short, almost imperceptible step more, giving rise to sterility, or reversing the tendency[.] Very likely you are on the right track; but you have something to do yet in that department. . . .

I am free to say that I never learned so much from one book as I have from yours. There remain 1000 things I long to say about it.—

Ever Yours | Asa Gray.

[1] Asa Gray, 'Review of Darwin's theory on the origin of species by means of natural selection', *American Journal of Science and Arts* 2d ser. 29 (1860): 153–84.

To Charles Lyell 10 April [1860]

Down Bromley Kent.
Ap. 10th

My dear Lyell

Thank you much for your note of the 4th. I am very glad to hear that you are at Torquay. I shd. have amused myself earlier by writing to you; but I have had Hooker & Huxley staying here, & they have fully occupied my time; as a little of anything is a full dose for me.— … There has been a plethora of Reviews, & I am really quite sick of myself.— There is very long Review by

Carpenter in Med.-Chirurg. Review: *very good* & well-balanced but not brilliant.

Carpenter speaks of you in thoroughily proper terms. There is a *brilliant* review by Huxley, with capital hits; but I do not know that he much advances subject: I *think* I have convinced him that he has hardly allowed weight enough to the cases of varieties of plants being in some degree sterile.—

To diverge from Reviews[:] Asa Gray sends me from Wyman (who will write) a good case of all the pigs being black in the Everglades of Virginia; on asking about cause, it seems (I have got capital analogous cases) that when the *black* pigs eat a certain nut, their bones become red & they suffer to certain extent, but that the *white* pigs lose their hoofs & perish; "& we aid by *selection* for we kill most of the young white pigs" This was said by man who could hardly read.—

By the way it is a great blow to me that you cannot admit to potency of natural selection; the more I think of it, the less I doubt its power for great & small changes.—

I have just read the Edinburgh, which without doubt is by Owen. It is extremely malignant, clever & I fear will be very damaging. He is atrociously severe on Huxley's lecture, & very bitter against Hooker.[1] So we three *enjoyed* it together: not that I really enjoyed it, for it made me uncomfortable for one night; but I have got quite over it today. It requires much study to appreciate all the bitter spite of many of the remarks against me; indeed I did not discover all myself.— It scandalously misrepresents many parts. He misquotes some passages altering words within inverted commas. Makes me say that the dorsal vertebræ of pigeons vary & refers to page where the word dorsal does not appear. Sneers at my saying a certain organ is the branchiæ of Balanidæ; whilst in his own "Invertebrata" published before I published on cirripedes, he calls them organs without doubt branchiæ.—

It is painful to be hated in the intense degree with which Owen hates me.—

Now for a curious thing about my Book, & then I have done. In last Saturday Gardeners' Chronicle, a Mr Patrick Matthews publishes long extract from his work on "Naval Timber & Arboriculture" published in 1831, in which he briefly but completely anticipates the theory of Nat. Selection.— I have ordered the Book, as some few passages are rather obscure but it, is certainly, I think, a complete but not developed anticipation! Erasmus[2] always said that surely this would be shown to be the case someday. Anyhow one may be excused in not having discovered the fact in a work on "Naval Timber". . . .

Farewell my dear Lyell | Yours affect | C. Darwin . . .

[1] Richard Owen's review of *Origin* and other works in the *Edinburgh Review* 111: 487–532 attacked, in addition to *Origin*, Huxley's lecture 'On species and races, and their origin', delivered at the Royal Institution of Great Britain, and Hooker's *Introductory essay to the Flora of Tasmania* (London, 1859), in which Hooker first announced his agreement with CD's theory.

[2] Erasmus Alvey Darwin was CD's elder brother.

[Following the publication of *Origin*, a Scottish gentleman-farmer, Patrick Matthew, claimed that he had anticipated CD's concept of natural selection in a publication in 1831 (*On naval timber and arboriculture; with critical notes on authors who have recently treated the subject of planting* (London and Edinburgh, 1831)). CD discussed Matthew's views in the 'Historical sketch' prefacing the third edition of *Origin*.]

To *Gardeners' Chronicle* [13 April 1860]

I have been much interested by Mr. Patrick Matthew's communication in the Number of your Paper, dated April 7th. I freely acknowledge that Mr. Matthew has anticipated by many years the explanation which I have offered of the origin of species, under the name of natural selection. I think that no one will feel surprised that neither I, nor apparently any other naturalist, had heard of Mr. Matthew's views, considering how briefly they are given, and that they appeared in the appendix to a work on Naval Timber and Arboriculture. I can do no more than offer my apologies to Mr. Matthew for my entire ignorance of his publication. If another edition of my work is called for, I will insert a notice to the foregoing effect.

Charles Darwin, Down, Bromley, Kent.

To Charles Lyell 15 April [1860]

Down Bromley Kent
Ap. 15th

My dear Lyell

... I was particularly glad to hear what you thought about not noticing Owen's Review. Hooker & Huxley thought it a sort of duty to point out alterations of quoted citations; & there is truth in this remark, but I so hated the thought, that I resolved not to do so. . . .

I must say one more word about our quasi-theological controversy about natural Selection, & let me have your opinion when we meet in London.— Do you consider that the successive variations in size of the crop of the Pouter Pigeon, which man has accumulated to please his caprice, have been due to "the creative & sustaining powers of Brahma". In the sense that an omnipotent & omniscient Deity must order & know everything, this must be admitted; yet in honest truth I can hardly admit it. It seems preposterous that a maker of Universes shd care about the crop of a Pigeon solely to please men's silly fancies. But if you agree with me in thinking such an interposition of the Deity uncalled for, I can see no reason whatever for believing in such interpositions in the case of natural beings, in which strange & admirable peculiarities have been naturally selected for the creature's own benefit. Imagine a Pouter in a state of nature wading into the water & then being buoyed up by its inflated crop, sailing about in search of food. What admiration this would have excited,—adaptation to laws of hydrostatic pressure &c &c &c.—

For the life of me I cannot see any difficulty in Natural selection producing the most exquisite structure, *if such structure can be arrived at by gradation*; & I know from experience how hard it is to name any structure towards which at least some gradations are not known.—

Ever yours | C. Darwin

The conclusion at which I have come, as I have told Asa Gray, is that such question, as is touched on in this note, is beyond the human intellect, like "predestination & free will" or "the origin of evil".

[In June 1858, while CD was writing on his theory of natural selection for eventual publication, he received a letter from Wallace containing a manuscript describing his theory of natural selection, which he had arrived at independently. Lyell and Hooker arranged for it to be published jointly with a brief account of his theory that CD had originally written to Asa Gray in 1857.]

To Alfred Russel Wallace 18 May 1860

Down Bromley Kent
May 18th 1860

My dear Mr Wallace

I received this morning your letter from Amboyna dated Feb. 16th, containing some remarks & your too high approbation of my book.[1] Your letter has pleased me very much, & I most completely agree with you on the parts which are strongest & which are weakest. The imperfection of Geolog. Record is, as you say, the weakest of all; but yet I am pleased to find that there are almost more Geological converts than of pursuers of other branches of natural science. . . .

I think geologists are more converted than simple naturalists because more accustomed to reasoning. Before telling you about progress of opinion on subject, you must let me say how I admire the generous manner in which you speak of my Book: most persons would in your position have felt some envy or jealousy. How nobly free you seem to be of this common failing of mankind.— But you speak far too modestly of yourself;—you would, if you had had my leisure done the work just as well, perhaps better, than I have done it.— . . .

The attacks have been heavy & incessant of late. . . . But I am got case-hardened, & all these attacks will make me only more determinately fight. Agassiz sends me personal civil messages but incessantly attacks me; but Asa Gray fights like a hero in defence.— .— Lyell keeps as firm as a tower, & this autumn will publish on Geological History of Man, & will there declare his conversion, which now is universally known.— I hope that you have received Hooker's splendid Essay.—[2] So far is bigotry carried, that I can name 3 Botanists who will not even read Hooker's Essay!! . . .

I am at work at my larger work which I shall publish in separate volumes.—[3] But from ill-health & swarms of letters, I get on very

very slowly.— I hope that I shall not have wearied you with these details.—

With sincere thanks for your letter, & with most deeply-felt wishes for your success in science & in every way believe me, | Your sincere well-wisher | C. Darwin

[1] Wallace's letter has not been found.
[2] Joseph Dalton Hooker, *Flora Tasmaniæ* (London, 1860).
[3] CD at this time planned to publish three volumes devoted to providing fuller evidence for the theory presented in *Origin*. In the event, only one of the planned works, *Variation*, was published.

To Asa Gray 22 May [1860]

Down Bromley Kent
May 22[d]

My dear Gray.

... Judging from letters ... & from remarks, the most serious omission in my book was not explaining how it is, as I believe, that all forms do not necessarily advance,—how there can now be *simple* organisms still existing.— ...

With respect to the theological view of the question; this is always painful to me.— I am bewildered.— I had no intention to write atheistically. But I own that I cannot see, as plainly as others do, & as I sh[d] wish to do, evidence of design & beneficence on all sides of us. There seems to me too much misery in the world. I cannot persuade myself that a beneficent & omnipotent God would have designedly created the Ichneumonidæ with the express intention of their feeding within the living bodies of caterpillars, or that a cat should play with mice. Not believing this, I see no necessity in the belief that the eye was expressly designed. On the other hand I cannot anyhow be contented to view this wonderful universe & especially the nature of man, & to conclude that everything is the result of brute force. I am inclined to look at everything as resulting from designed laws, with the details, whether good or bad, left to the working out of what we may call chance. Not that this notion *at all* satisfies me. I feel most deeply that the whole subject is too profound for the human intellect. A dog might as well speculate on the mind of Newton.— Let each man hope & believe what he can.—

Certainly I agree with you that my views are not at all necessarily atheistical. The lightning kills a man, whether a good one or bad one, owing to the excessively complex action of natural laws,—a child (who may turn out an idiot) is born by action of even more complex laws,—and I can see no reason, why a man, or other animal, may not have been aboriginally produced by other laws; & that all these laws may have been expressly designed by an omniscient Creator, who foresaw every future event & consequence. But the more I think the more bewildered I become; as indeed I have probably shown by this letter.

Most deeply do I feel your generous kindness & interest.—

Yours sincerely & cordially | Charles Darwin

[Hooker was at the meeting of the British Association for the Advancement of Science at Oxford. His letter is a report on the famous 'debate' with the bishop of Oxford, Samuel Wilberforce.]

From Joseph Dalton Hooker 2 July 1860

Botanic Gardens Oxford
July 2/60

Dear Darwin

I have just come in from my last moonlight saunter at Oxford . . . & cannot go to bed without inditing a few lines to you my dear old Darwin. I came here on Thursday afternoon & immediately fell into a lengthened revirie: without you & my wife I was as dull as ditch water & crept about the once familiar streets feeling like a fish out of water— I swore I would not go near a Section & did not for two days—but amused myself with the Colleges buildings & alternate sleeps in the sleepy gardens & rejoiced in my indolence. Huxley & Owen had had a furious battle over Darwins absent body at Section D.,[1] before my arrival,—of which more anon. H. was triumphant— You & your book forthwith became the topics of the day, & I d—d the days & double d—d the topics too, & like a craven felt bored out of my life by being woke out of my reveries to become referee on Natural Selection &c &c

&c— On Saturday I walked with my old friend of the Erebus Capt Dayman to the Sections & swore as usual I would not go in; but getting equally bored of doing nothing I did. A paper of a yankee donkey called Draper on "civilization according to the Darwinian hypothesis" or some such title was being read, & it did not mend my temper; for of all the flatulent stuff and all the self sufficient stuffers—these were the greatest, ... however hearing that Soapy Sam [Wilberforce] was to answer I waited to hear the end. The meeting was so large that they had adjourned to the Library which was crammed with between 700 & 1000 people, for all the world was there to hear Sam Oxon— Well Sam Oxon got up & spouted for half an hour with inimitable spirit uglyness & emptyness & unfairness, I saw he was coached up by Owen & knew nothing & he said not a syllable but what was in the Reviews— he ridiculed you badly & Huxley savagely— Huxley answered admirably & turned the tables, but he could not throw his voice over so large an assembly, nor command the audience; & he did not allude to *Sam's* weak points nor put the matter in a form or way that carried the audience. The battle waxed hot. Lady Brewster fainted, the excitement increased as others spoke—my blood boiled, I felt myself a dastard; now I saw my advantage—I swore to myself I would smite that Amalekite Sam hip & thigh if my heart jumped out of my mouth & I handed my name up to the President ([J. S.] Henslow) as ready to throw down the gauntlet— I must tell you that Henslow as president would have none speak but those who had *arguments* to use, & 4 persons had been burked by the audience & President for mere declamation: it moreover became necessary for each speaker to mount the platform & so there I was cocked up with Sam at my right elbow, & there & then I smashed him amid rounds of aplause— I hit him in the wind at the first shot in 10 words taken from his own ugly mouth—& then proceeded to demonstrate in as few more 1 that he could never have read your book & 2 that he was absolutely ignorant of the rudiments of Bot. Science— I said a few more on the subject of my own experience, & conversion & wound up with a very few observations on the relative position of the old & new hypotheses, & with some words of caution to the audience— Sam was shut up—had not one word to say in reply & the meeting *was dissolved forthwith* leaving you master of the field after 4 hours battle. Huxley who had borne all the previous brunt of the battle & who never before (thank God) praised me to my face, told me it was splendid, & that he did not know before what stuff I was made of— I have been congratulated & thanked by the blackest coats & whitest stocks in Oxford (for they hate their Bishop quite *[section illeg]* love) & plenty of ladies too have flattered me—but eheu & alas never is[2]

[1] Section D was the botanical and zoological section of the BAAS meeting.

[2] The passage '(for they ... never is' has been heavily inked over, and the rest of the letter is missing.

To T. H. Huxley 3 July [1860]

Sudbrook Park | Richmond | (I return to Down on Saturday)

July 3^{d}

My dear Huxley ...

I had letter from Oxford written by Hooker late on Sunday night, giving me some account of the awful battles which have raged about "species" at Oxford. He tells me you fought nobly with Owen, (but I have heard no particulars) & that you answered the B. of O. capitally.— I often think that my friends (& you far beyond others) have good cause to hate me, for having stirred up so much mud, & led them into so much odious trouble.— If I had been a friend of myself, I should have hated me. (how to make that sentence good English I know not.) But remember if I had not stirred up the mud some one else certainly soon would.— I honour your pluck; I would as soon have died as tried to answer the Bishop in such an assembly. Was Owen very blackguard? H. says that the Bishop turned me into ridicule & was very savage against you.— I hardly like to ask you to write, for I know how you are overworked; but I sh^{d} rather like to hear a bit about the battle.[1] I did not imagine that it would have turned up at Oxford; but I am now glad that I had no choice about going for I was utterly unfit.— The world surely will soon get weary of subject & let us have some peace. Though, on other hand, I do believe this row is best thing for subject.— As I am never weary of saying I sh^{d} have been **utterly** smashed had it not been for you & three others. . . .

I fancy from what Hooker says he must have answered the Bishop well.— God knows, I honour & thank you both.

Ever yours | C.D.

[1] Huxley's reply has been lost. The precise wording of his response to Wilberforce has never been established. In one version of the story, the bishop had asked Huxley whether he preferred to be descended from an ape on his grandfather's or his grandmother's side. Huxley gave his own account of the proceedings in a letter to a friend about two months later: 'If then, said I, the question is put to me would I rather have a miserable ape for a grandfather or a man highly endowed by nature and possessed of great means and influence and yet who employs those faculties for the mere purpose of introducing ridicule into a grave scientific discussion—I unhesitantly affirm my preference for the ape' (J. V. Jensen, 'Return to the Wilberforce–Huxley debate', *British Journal for the History of Science* 21 (1988): 168).

To John Medows Rodwell 15 October [1860]

15 Marine Parade | Eastbourne
Oct. 15th.

Dear Sir

I am truly obliged for your interesting letter. I am very far from being surprised at anyone not accepting my conclusions on the origin of species; as the argument rests almost solely on the view explaining & grouping phenomena, otherwise inexplicable. I have some confidence that I am in the main right; for I find it as yet a universal rule, that those naturalists who go a little way with me, the more they reflect on the subject the further they go.— I am at work on my larger work, but ill-health & other interruptions make my progress very slow.— Your remarks on language seem to me very striking; & the 'struggle for life' with words is quite new to me.— I had often thought that a striking resemblance might be traced in the genesis of words & species; but was much too ignorant to attempt it.— It was done to a certain extent some 4 or 6 months ago in the Cornhill Magazine by Lewis in one of his Zoological papers.—[1] Could you not publish an essay on the subject?

I have been particularly interested by your case of the Horses. I have somewhere read a nearly parallel case. I am sorry to give you trouble; but I sh^{d} very much like to know whether the case was published; & if you can give me any further particulars; such as how many Horses were affected; how soon they recovered &c.— Prof. Wyman has sent me an analogous case with respect to pigs in Florida: he was surprised at seeing them all black; & he found that they eat a certain root, which injures & kills the white pigs, but does not hurt the black; & the farmers added, "we help it by *selection*, for we kill the young white pigs".—

It is not white cats, but white cats with *blue* eyes, which are deaf,—if one eye is blue the cat is deaf on that one side—if your cat has *distinctly* blue eyes & is not deaf, I sh^{d} be particularly obliged for the fact, as it will be the sole exception which I have heard of.—

Pray accept my best thanks for your interesting letter & with respect, I beg leave to remain | Dear Sir | yours faithfully & obliged | Charles Darwin | of Down, Bromley, Kent

I shall be here for about 10 days

[1] George Henry Lewes, 'Studies in animal life', *Cornhill Magazine* 1 (1860): 61–74, 198– 207, 283–95, 438–47, 598–607, 682–90.

[Gideon Lincecum, an American physician and self-taught naturalist, on reading *Origin* sent this letter of observations on ants, which he considered to be evidence supporting CD's theory. CD sent the letter to the Linnean Society of London, which published an abstract of the letter edited by George Busk.]

From Gideon Lincecum 29 December 1860

Long Point. Texas.
29. Decr 1860.

Charles Darwin. M.A., | Down, Bromley, Kent. England.
Dear Sir. . . .

The species of Formica, which I have named Agricultural; is a large brownish red ant, dwells in paved cities, is a farmer, thrifty and healthy; is delligent and thoughtful, making suitable and timely arrangements for the changing seasons; in short, he is endowed with capacities sufficient to contend with much skill and ingenuity; and untiring patience, with the varying exigencies with which he may encounter in the life conflict.

When [he] selects a situation upon which to locate a city, if it is on ordinarily dry land, he bores a hole, around which he elevates the surface three, sometimes six inches; forming a low, circular mound; with a very gentle inclination from the center to its outer limits; which on an average is three to four feet from the entrance. But if the location is made on a low, or flat wet land, liable to inundation, though the ground may be perfectly dry at the time he does the work, he nevertheless elevates his mound in the form of a pretty sharp cone, to the hight of fifteen to twenty inches, sometimes even more, having the entrance near the apex. Around this, and its the same case with the upland cities, he clears the ground of all obstructions, levels and smoothes the surface to the distance of three or four feet from the gate of the city, giving it the appearance of a handsom pavement, as it really is. Upon this pavement not a spine of any green thing is permitted to grow, except a single species of grain bearning grass. Having planted it in a circle around, and two or three feet from the center of the mound, he nurses and cultivates it with constant care cutting away all other grasses and weeds that may spring up amongst it, and all round outside of the farm circle to the extent of one or two feet. The cultivated grass grows luxuriently, producing a heavy crop of small, white, flinty seeds, which under the microscope very much resembles the rice of commerce. When it gets ripe it is carefully harvested, and carried by the workers, chaff and all into the grainery cells, where it is divested of the chaff and packed away; the chaff is taken out and thrown beyond the limits of the pavement.

During protracted spells of wet weather, it sometimes happens that their provision stores become damp, and liable, as they are invariably seeds of some kind, to sprout and spoil. If this has occurred, the first fair day after the rain, they bring out the damp and damaged stores expose them to the sun till they are dry, when they carry back and pack away all the sound seeds, leaving all that are sprouted to waste. . . .

Most Respectfully. | Gideon Lincecum.

1861

To J. D. Hooker 4 February [1861]

Down Bromley Kent
Feb 4th

My dear Hooker

I was delighted to get your long, chatty letter & to hear that you are thawing towards science. I almost wish you had remained frozen rather longer; but do not thaw too quickly & strongly. No one can work long like you used to do. Be idle; but I am a pretty man to preach, for I cannot be idle, much as I wish it, & am never comfortable except when at work. The word Holiday is written in a dead language for me, & much I grieve at it.—

We thank you sincerely for your kind sympathy about poor Etty,[1] who about a fortnight ago had three terrible days of sickness & was given loads of calomel, which I always dread. She has now come up to her old point & can sometimes get up for an hour or two twice a day.— Poor George[2] has literally every tooth in his head, except a few lower incisors decayed: they have all gone suddenly together & been stopped & drawn by Mr Woodhouse & I fear this points to some deep flaw in his constitution, which was formerly indicated by his intermittent pulse. Never to look to the future, or as little as possible, is becoming our rule of life.— What a different thing life was in youth, with no dread in the future;—all golden, if baseless, hopes.— …

I have been doing little, except finishing the new [3rd] Edit. of Origin, & crawling on most slowly with my volume on "Variation under Domestication,"—& how much to give under each head puzzles me dreadfully.— … I long to be at Drosera[3] again: I cannot persuade myself that it is the weight of $\frac{1}{78,000}$ of a grain of solid substances which causes such plain movement; nor that it is in most of the cases the chemical nature; & what it is, stumps me quite.

Do you remember a tall Silver Fir in the field in front of this House. It was so ugly we have grubbed it up. The hole was 3 ft. 6 inches in diameter with touching roots extending much further. Rings of growth 110.— It seems to have been planted on little mound of made earth; & I have got several pounds of this earth from exactly under centre in a sort of cone, where it is hardly possible seeds could have got for last 60 or 80 years. I have saved this earth on purpose to spite you about seeds germinating. And I have a great advantage, for if any come up in my study it will be good case; if none do (as I rather fear) then I shall say there were no seeds in earth!

I had a long letter about a week ago from Asa Gray, but I did not send it, thinking you would not care for it, as it almost wholly is on Design & quasi theological. He tells me that two of my opponents are gone almost demented: Bowen denying that any deviation is ever inherited; & [Louis] Agassiz maintaining that Greek Latin & Sanscrit are not affiliated but, like the races of men, are autochthonous! It is impossible to argue better for us.— . . .

I am glad to hear so good an account of your Willy.[4] A lull after a gang of children is very pleasant.

Ever yours affectionately | C. Darwin

1 Henrietta Emma Darwin, CD's eldest daughter.
2 George Howard Darwin, CD's second son.
3 The common sundew, an insectivorous plant.
4 William Henslow Hooker, Hooker's eldest son.

To Leonard Horner 14 February [1861]

Down Bromley Kent
Feb. 14th

My dear Mr. Horner

I must just thank you for your note, but I will take advantage of your kind & considerate offer of discussing the points referred to till we meet. . . .

Man does not cause any variation he only accumulates any which occur: I do not suppose that God intentionally gave the parent Rock-Pigeon a tendency to vary in size of Crop, so that man by selecting such variations should make a Pouter; so under nature, I believe variations arise, as we must call them, in our ignorance accidentally or spontaneously, & these are naturally selected or preserved from being beneficial to the successive individual animals in their struggles for Life.— I know not whether I make myself clear.—

Believe me | My dear Mr Horner | Yours very sincerely | C. Darwin

To Asa Gray 17 February [1861]

Down Bromley Kent
Feb 17th.—

My dear Gray

I received your note of Feb 5th this morning. . . . I am glad to hear of the 250 copies now I presume at Trübners;[1] I wrote to him today about a few advertisements &c &c & to send me copies for distribution. I hope for my sake, as well as yours, that Murray & Lyell will not prove entirely true, that it is impossible to circulate a pamphlet in England.—

I fear that the state of the U. States must stop all interest in everything not political.— . . . The Printers have been very slow with my new Edit. of Origin so that I have been able to insert notice of your Pamphlet with title in full, which I am especially glad of.— . . .

. . . You ask about Drosera: if you like to try anything, put the minutest atom (under a lens) on point of fine needle on any one single extreme marginal gland of a leaf, which has all the hairs equally expanded & watch it or look again in 10 minutes.— Or put fragment of Hair of your head & look in a hour's time. I intend trying many more experiments this summer & then publishing: I am doubtful on many points. But the worst is that my health is failing much. I literally cannot listen to a novel for ½ hour without fatigue. My good dear wife declares, I must go with our whole family (if my girl can be moved) for two months to Water Cure; & I fear I must, but it will be ruin to all my experiments.— I remember formerly having read your extremely curious observations on Tendrils, but I thank you for telling me of them.—

With respect to Design &c you say that you suppose that I have "not brought forward my real objections against your views.—"

I have no real objection, nor any real foundation, nor any clear view.— As I before said I flounder hopelessly in the mud.—

You have amused me much by your account of Agassiz's denying the community of descent of allied languages, & of Bowen denying heredity. I cannot believe that Bowen is a strong man. What an odd & foolish fancy he must think it that all breeders of Race-Horses, Cattle & pigs &c should keep pedigrees, & would certainly prefer breeding from a poor animal of a good pedigree than from the finest of bad pedigree.— These men in fact work on **my** (I wish I dared say **our**) side. . . .

Most cordially & gratefully yours | Charles Darwin . . .

[1] CD had arranged to reprint Asa Gray's three articles on *Origin* from the *Atlantic Monthly* (6: 109–16, 229–39, 406–25) under the title *Natural selection not inconsistent with natural theology* (London, 1861).

From Henry Walter Bates 18 March 1861

King St Leicester
18 March 1861

Dear Sir

At last the Ent. Society have printed my paper & I am enabled to send you ⟨a⟩ copy according to promise.[1] . . .

I think there are about 3 points of interest arising from the review of the species of Papilio— . . . These are

1 The derivation of the Amazonian fauna. I confess I was not prepared for the result to which I was obliged to arrive after a close examination of the species & their distribution—viz: that the Guiana region must ⟨ha⟩ve been the seat of an ancient & peculiar fauna transmitted through vast lapses of time; & that thence was derived the fauna of the Amazon valley.— Also that it was still so rich in endemic species. Surely I am right in deriving the conclusion that there can have been no great extinction here ⟨d⟩uring the glacial epoch.[2]

2 The widely different variability of species when under different local conditions in localities widely apart— . . .

3 The permanency of local varieties after they have become established. It is still the favourite argument of our best naturalists that varieties will always return to their normal form & that they

will inter breed & produce fertile offspring. This argument is derived from the observations of varieties produced by domesti⟨cation⟩—a false guide—such varieties are too rapidly made, to be compared with the slow alterations of the whole organism which takes place in nature & affects, I have no doubt, at length the reproductive elements.— In the genus Papilio there is a set of local varieties all connected by fine gradations of differences; & yet in one well established case two of these varieties exist in contact & do not show the slightest tendency to amalgamate.— It is a case exactly parallel to what would be if we were to find in the wild state a series of graduated local varieties between the horse & ass;— I thought it likely I should find in the Natural History of the Horse & ass some data to prove the parallel & turned to a paper by Blyth lately published on the varieties of wild ass.[3] I was surprised to find how little was satisfactorily known on the subject & how uncertain & vacillating is the state of our knowledge of the species & varieties of these conspicuous animals. . . .

I hope you will excuse this rather rambling letter & favour me with your opinion on my pap⟨er⟩ at your earliest convenience

Yours sincerely | H W Bates

[1] H. W. Bates, 'Contributions to an insect fauna of the Amazon valley. Diurnal Lepidoptera', *Transactions of the Entomological Society of London* n.s. 5 (1858–61): 223–8, 335–61.

[2] In his paper, Bates put forward the view that the *Papilio* species peculiar to the Amazon region were modifications of species from nearby regions such as Guiana. Bates claimed that if the glacial period had extended into equatorial regions, as CD had argued, it would have resulted in the extinction of many forms peculiar to those regions.

[3] Edward Blyth, On the different animals known as wild asses, *Journal of the Asiatic Society of Bengal* 28 (1859): 229–53. (Reprinted in *Annals and Magazine of Natural History* 3d ser. 6 (1860): 233–54.)

To William Bernhard Tegetmeier 22 March [1861]

Down Bromley Kent
March 22d

My dear Sir

I ought to have answered your last note sooner; but I have been very busy. How wonderfully successful you have been in breeding Pouters! you have good right to be proud of your accuracy of eye & judgment. I am in the thick of Poultry;[1] having just commenced

& shall be truly grateful for the skulls, if you can send them by any conveyance to the Nag's Head next Thursday. . . .

You ask about my crossed Fowls; & this leads me to make a proposition to you, which I hope cannot be offensive to you.— I trust you know me too well to think that I would propose anything objectionable to the best of my judgment. The case is this. For my object of treating Poultry, I must give sketch of several breeds with remarks on various points. I do not feel strong on subject. Now when my M.S. is fairly copied in an excellent hand-writing; would you read it over, which would take you at most an hour or two, & make comments in pencil on it; & accept, like a Barrister, a fee, we will say of a couple of guineas.— This would be a great assistance to me; specially if you would allow me to put note, stating that you, a distinguished judge & Fancier, had read it over. I would state that you doubted or concurred as each case might be: of course striking out what you were sure was incorrect. There would be little new in my M.S. to you; but if by chance you used any of my facts or conclusions, before I published, I sh^{d} wish you to state that they were on my authority; otherwise I shall be accused of stealing from you.— There will be little new except that perhaps I have consulted some out of the way books, & have corresponded with some good authorities.

Tell me frankly what you think of this; but unless you will oblige me by accepting remuneration, I cannot & will not give you such trouble.—[2] I have little [doubt] that several points will arise which will require investigation, as I care for many points disregarded by Fanciers; & according to any time thus spent you will, I trust, allow me to make remuneration. I hope that you will grant me this favour.—

There is one assistance which I will *now* venture to beg of you, viz to get me, if you can, another specimen of *old white* Angora Rabbit. I want it dead for Skeleton; & *not* knocked on Head. Secondly I see in Cottage Gardener (March 19th. p. 375) there are (impure) Half-lops with one ear quite upright & shorter than the other lopped ear. I much want a dead one.— Baker cannot get one.— Baily is looking out; but I want two specimens. Can you assist me, if you meet any Rabbit fancier? I have had rabbit with one ear more lopped than the other; but I want one with one ear quite upright & shorter, & the other quite long & lopped.—

My dear Sir | Yours sincerely | Ch. Darwin

[1] CD had begun the chapter on poultry for *Variation*.
[2] An entry in CD's Account books (Down House MS) records a payment to Tegetmeier of £5 5*s*. for 'scientific assistance'.

To H. W. Bates 26 March [1861]

Down Bromley Kent
March 26

Dear Sir

I have read your papers with extreme interest & I have carefully read every word of them. They seem to me to be far richer in facts on variation, & especially on the distribution of varieties & subspecies, than anything which I have read. Hereafter I shall reread them, & hope in my future work to profit by them & make use of them. The amount of variation has much surprised me. The analogous variation of distinct species in the same regions strikes me as particularly curious. The greater variability of female sex is new to me. Your Guiana case seems in some degree analogous, as far as plants are concerned, with the modern plains of La Plata, which seem to have been colonised from the north, but the species have been hardly modified.—

I have been particularly struck with your remarks on the Glacial period. You seem to me to have put the case with admirable clearness & with crushing force. I am quite staggered with the blow & do not know what to think. Of late several facts have turned up leading me to believe more firmly that the Glacial period did affect the Equatorial Regions; but I can make no answer to your argument; & am completely in a cleft stick. By an odd chance I had only a few days ago been discussing this subject, in relation to plants, with D^r. Hooker, who believes to a certain extent; but strongly urged the little apparent extinctions in the Equatorial regions. I stated in a letter some days ago to him, that the Tropics of S. America seem to have suffered less than the Old World.

There are many perplexing points, temperate plants seem to have migrated far more than animals— Possibly species may have been formed more rapidly within Tropics than one would have expected. I freely confess that you have confounded me: but I cannot yet give up my belief that the Glacial period did to certain extent affect the Tropics.—

Would you kindly answer me 2 or 3 questions if in your power. — When species (A) becomes modified in another region into a well marked form (C), but is connected with it by one (or more) gradational form (B,) inhabiting an intermediate region; does this form (B) generally exist in equal numbers with (A) & (C), *or inhabit an equally large area*?— The probability is that you cannot answer this question: though one of your cases seem to bear on it.— . . .

In Butterflies, in which the sexes are differently coloured, is the male or female most beautiful in *our* eyes? Do you know in Tropics any strictly nocturnal moths with gaudy colours? As with Birds, have you ever noticed that female butterflies make any selection of the male with which they copulate? Do several males pursue same female? Are butterflies attracted by gay colours, as it has been asserted Dragon-flies are. Any authentic facts on the courtship of Butterflies would be *most thankfully* received & quoted by me. But I can see how very improbable it is that anything sh[d]. have been observed.—

You will, I think, be glad to hear that I now often hear of naturalists accepting my views more or less fully: but some are curiously cautious in running risk of any small odium in expressing their belief.

With cordial thanks & respect | Believe me my dear Sir | Yours sincerely | C. Darwin

(Have you received copy of new Edit. of Origin?)

From H. W. Bates 28 March 1861

King St Leicester
28 March 1861

My Dear Sir

I received your kind letter today . . .

With regard to the question you ask me—whether an intermediate local form B is numerous & widely dispersed between the ranges of its extreme forms A & C: the facts that I have on the subject are numerous & rather ill digested at present. They are complicated in themselves & difficult. I know many instances of two local forms separated by a wide space without apparent natural barriers, untenanted by intermediate forms. There are many others (closely allied) which exist together on their mutual frontiers without blending. There are others which are very polymorphic in a central region, whilst in other localities E. W. N. & S. segregated into several well defined local varieties or admitted species. I will mention now one case only that meets your question. It is one, whose correctness I am quite sure of. On the dry soils, supporting for the most part a thinner forest growth, of the hilly regions of Guiana & Venezuela there exists a conspicuous Heliconia, H. Melpomene, it is abundant also in the central parts of the lower Amazons where the Hilly sandy country occurs & is found on the N. & S. Shores in extreme profusion. In the moist alluvial plains eastward to Pará & westward to Peru & Bolivia, not a single individual is to be seen. In its place, occupying exactly the same sphere as it were, is H. Thelxiope a form so strikingly peculiar in colours that no one has ever doubted it to be a perfectly distinct species; it swarms in individuals & is nearly constant in its specific characters. Now wherever this form comes into contact with Melpomene there exists a number of intermediate varieties many of which have been described as species: they *are rare*, & very *restricted in range*. They have puzzled Lepidopterists & it has been almost settled that they were hybrids. I am convinced they are not *hybrids*; I never saw Melpomene & Thelxiope in copula; and besides these pretended hybrids exist at Demerara & Cayenne, where Thelxiope *does not occur*. I am thoroughly convinced that Thelxiope is a local variety of Melpomene, having all the appearance of a species, but created by the influences to which it has been subjected, out of Melpomene; & that the intermediate forms are the gradations. I have found Melpomene in copula with these forms & perhaps it may sometimes be so with Thelxiope, but that would not affect the case much I think.

The group from which I have supplied this example is very interesting,—being restricted to S. America, which swarms with its species; & apparently a modern creation. I am now studying it with a view to writing upon it & I think I have got a glimpse into the laboratory where Nature manufactures her new species.

—The other question you ask me relates to sexual selection.— I *have* seen Papilios attracted by bright colours in the forest, namely by the scarlet sepals of certain plants having inconspicuous or no flowers. I have also seen repeatedly many males following a female.— On this subject I will point out to you the following facts. In the æneas section of the genus Papilio, the males are generally of extremely brilliant colours, velvety black vivid green, carmine, & red with opalescent reflections,—the females are plainer & so different from the males that they were generally held to be distinct species until I took them in copulâ. Now it is not *all* the species that present this disparity. In one P. Panthonus the two sexes are exactly alike in colours, the male only being a trifle brighter; from this species the divergence may be traced getting wider & wider from species to species up to P. Sesostris or P. Childrenæ, where the dissimilarity reaches its highest point. There is exactly the same phenomenon presented in the genus Epicalia & others. The females, however, vary in localities;—& although I thoroughly believe in your theory of sexual selection yet I think that local circumstances have some effect on colours. I do not however think that sunlight is the direct cause of vivid colours, although beautifully coloured male butterflies are almost always in the sun (genus Catagramma) whilst their drab partners are in the shade. I think the causes lie in the abundance of food, warmth & moisture of the atmosphere & even in the sluggish state of the atmosphere because the brightest coloured butterflies are not generally found within the influence of the Atlantic sea breezes, but in the sultry valleys of the Andes & the centre of the continent & these causes operate on the larva & so by correlation on the perfect insect.

I cannot say that I have noticed female butterflies pointedly selecting their male partners. The extremely rich male butterflies of the genus Catagramma live & sport together in the sun light all day, whilst their plainer coloured females are confined to the shades of the forest. The males too appear to be very much more numerous than the females,— I think *there is no* doubt now about this significant fact   there must be many hundred males to one female of these butterflies, & in another genus Cybdelis where the males are also distinguished by rich colours the males are immensely more numerous than the females. In another beautiful genus, Megistanis no female has yet been found although the males are very numerous. In Callithea, however where the males are the most richly coloured of all butterflies, both sexes are found in equal numbers, but *here the females* are not *much* behind the *males* in *beauty*. The males which sport together all day in the sunlight disappear about 4–5 P.M. & I have watched them then flying off to the forest (towards the summits of the trees) where doubtless some of them find their mates.—

There are no strictly nocturnal Lepidoptera gaily coloured in the tropics: There is no scarcity of brilliantly col[d]. moths but they are all day fliers.

Perhaps the above facts will illustrate a little the problems of sexual disparity & beauty of colours. There is no phase of your theory I like better than its explanation of the subtle adaptations of organic beings. There is one topic which Entomology will help to illustrate viz. that of mimetic analogies. I have an immense number of facts on this subject. Some of these resemblances are perfectly staggering,—to me they are a source of constant wonder & thrilling delight. It seems to me as though I obtain a glimpse of an *intelligent motive* pervading nature, as well as of the mighty never-resting wonder working laws that regulate all things.— . . .

Believe me My Dear Sir Yours sincerely | H W Bates . . .

[John Stevens Henslow, professor of botany at Cambridge University, CD's teacher and friend during his Cambridge student years, and the man responsible for securing CD's place on board the *Beagle*, fell seriously ill in the spring of 1861 and died on 16 May. J. D. Hooker was Henslow's son-in-law.]

To J. D. Hooker 23 [April 1861]

Down
23[d].

My dear Hooker

I am much pained to think of poor dear Henslow's state. . . . I write now only to say that if Henslow, you thought, would really like to see me, I would of course start at once. The thought had once occurred to me to offer, & the sole reason why I did not was that the journey with the agitation would cause me probably to arrive utterly prostrated.

I sh[d]. be certain to have severe vomiting afterwards, but that would not much signify, but I doubt whether I could stand the agitation at the time. I never felt my weakness a greater evil. I have just had specimen for I spoke a few minutes at Linn[ean] Soc on Thursday & though extra well, it brought on 24 hours vomiting. I suppose there is some Inn at which I could stay, for I sh[d] not like to be in the House (even if you could hold me) as my retching is apt to be extremely loud.—

I sh[d]. never forgive myself, if I did not instantly come, if Henslow's wish to see me was more than a passing thought.

My dear old friend | Your affect | C. Darwin

P.S. Judge for me: I have stated exact truth: but remember that I sh[d]. *never* forgive myself, if I disappointed the most fleeting wish of my master & friend to whom I owe so much.—

To J. D. Hooker 23 [April 1861]

Down Bromley Kent
23[d].

My dear Hooker

I was very glad to get your letter this morning (to which I wrote brief answer), though it contained so melancholy an account of poor dear & honoured Henslow. How strange & pathetic an account you give of his mental state; & how his kind feelings shine out. He truly is a model to keep always before one's eyes. How I wish his sufferings were closed in the long & tranquil sleep of death. It must be very depressing to you, & I am glad you can read & think on other subjects.—

I quite agree with what you say on Lieut. Hutton's Review (who he is, I know not): it struck me as very original: he is one of the very few who see that the change of species cannot be directly proved & that the doctrine must sink or swim according as it groups & explains phenomena. It is really curious how few judge it in this way, which is clearly the right way. . . .

I dined with Bell at Linn[ean] Club, & liked my dinner: though it must be confessed they are rather a poor set of muffs, & I sat by the muffiest, viz Miers.— But dining out is such a novelty to me that I enjoyed it. Bell has a real good heart.— I liked Rolleston's paper. . . . I had a dim perception of the truth of your profound remark, that he wrote in fear & trembling "of God, man & monkeys", but I would alter it into God, man, Owen & monkeys.— Huxley's letter was truculent & I see that everyone thinks it too truculent; but in simple truth I am become quite demoniacal about Owen, worse than Huxley, & I told Huxley that I sh^{d} put myself under his care to be rendered milder.[1] But I mean to try to get more angelic in my feelings; yet I never shall forget his cordial shake of the hand when he was writing as spitefully as he possibly could against me. But I have always thought you have more cause than I to be demoniacally inclined towards him.— Bell told me that Owen says that the Editor mutilated his article in Edinburgh R[eview] & Bell seemed to think it was rendered more spiteful by Editor; perhaps the opposite view is as probable: Oh dear this does not look like becoming more angelic in my temper.

I had splendid long talk with Lyell (you may guess how splendid, for he was many times on his knees with elbows on sofa & rump high in air) on his work in France: he seems to have done capital work in making out age of the celt-bearing beds; but the case gets more & more complicated. All, however, tends to greater & greater antiquity of man. . . .

I shall be anxious for your next letter about Henslow.

Farewell with sincere sympathy | My old friend | C. Darwin

. . .

1 Huxley's letter, published in the *Athenæum*, 13 April 1861, was part of an exchange with Owen; in it, he accused Owen of 'very serious errors respecting matters of anatomical fact'.

To Thomas Davidson 30 April 1861

Down. Bromley. | Kent
April 30th 1861

My dear Sir

I thank you warmly for your letter. . . .

Pray do not think that I feel the least surprise at your demurring to a ready acceptance [of my theory]; in fact I should not much respect anyone's judgment who did so that is if I may judge others from the long time which it has taken me to go round. Each stage of belief cost me years. The difficulties are as you say many & very great: but the more I reflect the more they seem to me to be due to our underestimating our ignorance. I belong so much to old times that I find that I weigh the difficulties from the imperfection of the geological record, heavier than some of the younger men.— I find to my astonishment and joy that such good men as Ramsay,

Jukes[,] Geikie and one old worker Lyell do not think that I have in the least exaggerated the imperfection of the record. If my views ever are proved true our current Geological views will have to be considerably modified—

My greatest trouble is not being able to weigh the direct effects of the long continued action of changed conditions of life without any selection, with the action of selection on mere accidental (so to speak) variability— I oscillate much on this head but generally return to my belief that the direct action of the conditions of life has not been great. At least this direct action can have played an extremely small part in producing all the numberless and beautiful adaptations in every living creature.

With respect to persons belief, what does rather surprise me is that anyone (like Carpenter) should be willing *to go so very far* as to believe that all Birds may have descended from one parent, and not go a little farther and include all the members of the same great Division; for on such a scale of Belief, all the facts in Morphology and in Embryology (the most important in my opinion of all subjects) become mere Divine Mockeries. This leads me to remark how singularly few have judged the argument on right principles many complain that I have not proved that any one species changes into another and they ignore the fact that the view given, apparently groups together and explains many phenomena. No one urges as a fatal objection to the Theory of Light that the undulations in the ether cannot be proved—or the very existence of ether, yet because the undulatory theory explains much it is now universally admitted— ...

I am at present at work on "Variation under Domestication" and slow and laborious work I find it, but I think it will throw some little light on the laws of variation— ...

With cordial thanks. | Pray believe me | My dear Sir. | Yours very sincerely | Ch Darwin ...

To J. D. Hooker 18 [May 1861]

Down. | Bromley. | Kent. S.E.
18th

My dear Hooker

I was very glad to hear that poor dear Henslow is at rest. I fully believe a better man never walked this earth. What a loss he will be to his parish! I can well believe how you will miss him. I well remember his saying before you married that if he could have picked out anyone for his son-in-law, it would have been you.— How kind he was to me as an undergraduate constantly asking me to his House & taking me long walks. I am thankful to think that at the time I fully enjoyed & appreciated his kindness. I suppose Babington will be Professor at Cambridge. What a contrast!

Give our kindest remembrances to M^rs. Hooker. My wife admired from her heart poor dear Henslow

Farewell my dear Friend | Your affect— | C. Darwin ...

To John Frederick William Herschel 23 May [1861]

Down. | Bromley. | Kent. S.E.
May 23^d

Dear Sir John Herschel

You must permit me to have the pleasure to thank you for your kind present of your Physical Geography.[1] I feel honoured by your gift, & shall prize this Book with your autograph. I am pleased with your note on my book on species, though apparently you go but a little way with me.[2] The point which you raise on intelligent Design has perplexed me beyond measure; & has been ably discussed by Prof. Asa Gray, with whom I have had much correspondence on the subject.— I am in a complete jumble on the point. One cannot look at this Universe with all living productions & man without believing that all has been intelligently designed; yet when I look to each individual organism, I can see no evidence of this. For, I am not prepared to admit that God designed the feathers in the tail of the rock-pigeon to vary in a highly peculiar manner in order that man might select such variations & make a Fan-tail; & if this be not admitted (I know it would be admitted by many persons), then I cannot see design in the variations of structure in animals in a state of nature,—those variations which were useful to the animal being preserved & those useless or injurious being destroyed. But I ought to apologise for thus troubling you.—

You will think me very conceited when I say I feel quite easy about the ultimate success of my views, (with much error, as yet unseen by me, to be no doubt eliminated); & I feel this confidence, because I find so many young & middle-aged truly good workers in different branches, either partially or wholly accepting my views, because they find that they can thus group & understand many scattered facts. This has occurred with those who have chiefly or almost exclusively studied morphology, geographical Distribution, systematic Botany, simple geology & palæontology. Forgive me boasting, if you can; I do so because I sh^{d} value your partial acquiescence in my views, more than that of almost any other human being.—

Believe me with much respect | Yours, sincerely & obliged | Charles Darwin

[1] J. F. W. Herschel, 'Physical geography' (Edinburgh, 1861), a revised version of his essay in *Encyclopædia Britannica*, 8th edition (Edinburgh, 1853–60).

[2] CD had heard that Herschel had referred to natural selection as the 'law of higgledy-piggledy'. See *Correspondence* vol. 7, letter to Charles Lyell, [10 December 1859].

From Emma Darwin [June 1861]

I cannot tell you the compassion I have felt for all your sufferings for these weeks past that you have had so many drawbacks. Nor the gratitude I have felt for the cheerful & affectionate looks you have given me when I know you have been miserably uncomfortable.

My heart has often been too full to speak or take any notice I am sure you know I love you well enough to believe that I mind your sufferings nearly as much as I should my own & I find the only relief to my own mind is to take it as from God's hand, & to try to believe that all suffering & illness is meant to help us to exalt our minds & to look forward with hope to a future state. When I see your patience, deep compassion for others self command & above all gratitude for the smallest thing done to help you I cannot help longing that these precious feelings should be offered to Heaven for the sake of your daily happiness. But I find it difficult enough in my own case. I often think of the words "Thou shalt keep him in perfect peace whose mind is stayed on thee".[1] It is feeling & not reasoning that drives one to prayer. I feel presumptuous in writing thus to you.

I feel in my inmost heart your admirable qualities & feelings & all I would hope is that you might direct them upwards, as well as to one who values them above every thing in the world. I shall keep this by me till I feel cheerful & comfortable again about you but it has passed through my mind often lately so I thought I would write it partly to relieve my own mind.[2]

[1] Isaiah 26:3.
[2] CD wrote at the end of letter: 'God Bless you. C.D. June 1861'.

[In 1839 CD had published a paper on the famous geological puzzle of the parallel terraces or 'roads' of Glen Roy in the district of Lochaber, southern Inverness-shire, Scotland, arguing that they were of marine origin. CD had encouraged Jamieson to 'take up Glen Roy' earlier in 1861, and was eventually, though reluctantly, pursuaded by his arguments that the terraces were glacial features. In his *Autobiography*, p. 84, CD called his paper 'a great failure' and stated that he was 'ashamed of it', partly owing to his faulty scientific methodology:

> Having been deeply impressed with what I had seen of the elevation of the land in S. America, I attributed the parallel lines to the action of the sea; but I had to give up this view when Agassiz propounded his glacier-lake theory. Because no other explanation was possible under our then state of knowledge, I argued in favour of sea-action; and my error has been a good lesson to me never to trust in science to the principle of exclusion.]

From Thomas Francis Jamieson 3 September 1861

Dear Sir

I returned a few days ago from a trip to Lochaber where I spent a fortnight and now hasten to present you with some of the results of my visit, and I may at once state that all I saw tended to impress upon me the conviction that these parallel roads have been formed along the margin of freshwater lakes and finding the marks of ice action so plain over the whole district I cannot help thinking that Agassiz hit upon the true solution of the problem when he pronounced these marks to be the effect of glacier-lakes.

I attentively examined the entrance to Loch Treig, and found both sides of the gorge to present the clearest evidence of most intense glacial action, and that to heights of many hundred feet above the like-rounded rocks, scores, flutings & perched blocks abound and all these phenomena are most conclusively seen to have been effected by a great volume of ice flowing down the valley now occupied by the Lake, and issuing out by this gorge into Glen Spean. . . . Mr Milne says he could find no transverse scratches on the rocks of Glen Spean. I was however more fortunate and found them on the north side of the Spean opposite the entrance to Loch Treig, as Agassiz had correctly described. The profusion of blocks on this side of Glen Spean over an immense expanse is very striking—in fact the whole ground about the mouth of Loch Treig is a perfect study of glacial action. . . .

. . . In looking up Glen Roy from the Gap, I was much struck by the extreme neatness & precision of the lines which seemed to me very unlike what might be expected from the shore of a lake subject to tidal action. Very different from the appearance of true old coastlines which I had seen on the West coast of Argyleshire last summer. . . . But what seemed to me even more important evidence in this respect was the wonderfully fine preservation of the deltas at the mouth of some of the streams near the head of Glen Roy. These deltas have the appearance of being lodged in the waters of a placid lake, even in a stagnant pool, so undisturbed is the outline of some of them. This seems inexplicable to me had the lake been an arm of the sea, subject to the flux & reflux of the tides. . . .

. . . The glacial markings on the syenitic granite on the N. side of Glen Spean opposite the entrance to Loch Treig are amongst the finest specimens of ice-work I have seen, this with the heaps of moraine matter & the perfect wilderness of boulders made me stare with astonishment how any one, after Agassiz had drawn attention to all this, could go on the ground & yet deny that there had been any glacier here! I do not suppose there is any place in Britain where the traces of a great ice stream are more complete. . . .

I have written the above notes very hastily but shall be happy to answer any queries you may wish to put. I may mention that I had most villainous weather, which prevented me making so long excursions as I otherwise might have done; and it was only by going doggedly to work with a waterproof & umbrella that I could get any thing done at all.

Yours ever | Thos. F. Jamieson

Ellon, Aberdeens^{h}.
3^{d}. Sept. 1861.

To T. F. Jamieson 6 September [1861]

Down. | Bromley. | Kent. S.E.
Sept 6th

Dear Sir

I thank you sincerely for your long & very interesting letter. Your arguments seem to me conclusive. I give up the ghost. My paper is one long gigantic blunder.

I suppose & hope that you will publish an account of what you have observed. The case seems very interesting. What a wonderful record of the old icy lakes do these shores present! It really is a grand phenomenon. I have been for years anxious to know what was the truth, & now I shall rest contented, though ashamed of myself.— How rash it is in science to argue because any case is not one thing, it must be some second thing which happens to be known to the writer.—

I will take the liberty to forward your letter to Sir C. Lyell, as I am sure he would like to read it.—

With very sincere thanks. Pray believe me, my dear Sir | Yours sincerely | Ch. Darwin

Did I not say that you would be able to settle the question?—

To Asa Gray 16 September [1861]

Down Bromley Kent
Sept. 16th

My dear Gray

In the whirl of your public affairs, science may be forgotten, or if not forgotten you may have no inclination to write. But if so inclined I sh^{d}. be very glad to have a little information on any cases of dimorphism, like that of Primula, & to be allowed to quote you.— To make sure that you may understand what I want to know: I give this beautiful diagram of the two forms of Primula.

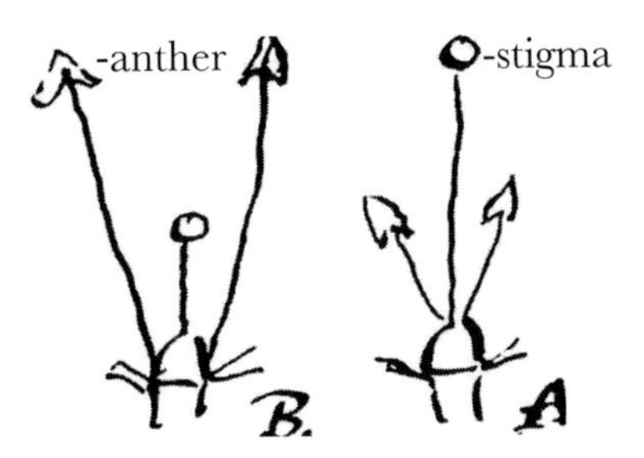

I think I have made out their good or meaning clearly. The pollen of A is fitted for stigma of B & conversely. The individuals are divided into two sets or bodies, like the males & females of Quadruped; but both in Primula are hermaphrodites: but I will not enter on details, as I will soon send a paper to Linn[ean] Soc.^y.— I should be *eminently* glad to know of other analogous cases. Are the two forms ever borne on same plant? Thyme is a different case: as the one form is simply female, stamens having aborted; these females, however, I find to be the most productive of seed. Some of the species of Linum offer this case:

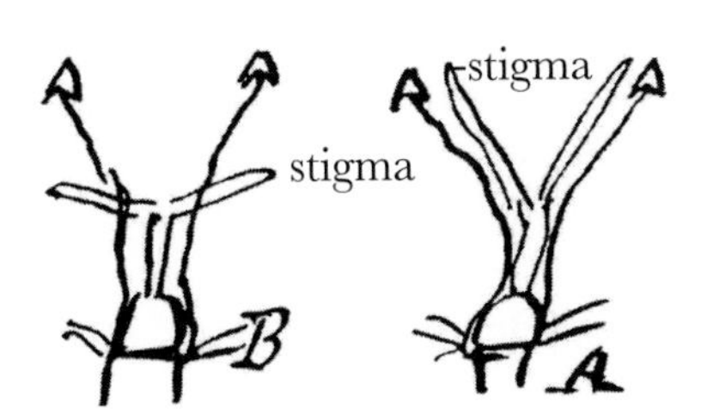

I am rather too late to experiment fairly; but I am almost sure that pollen of A is *absolutely* sterile on stigma of A: but good for stigma of B: whereas pollen of B is good for its own stigma B & for stigma of A.—

This subject interests me much, so do help me if you can; for I have some *very faint* hopes that it may throw some light on Hybridisation;[1] I have sown seeds of Primulas raised in very scanty numbers from stigmas fertilised by what I call a homomorphous union; ie by pollen from plant of same form.— …

I have almost finished my long Orchis paper, & then I must go back to my true work on Cocks & Hens, fowls & rabbits. Eheu Eheu, what much better fun observing is than writing.

By the way I have just been amusing myself in looking at Dionæa in aid of my Drosera work. How curious it is to see a fly caught & how beautiful are the adaptations compared with Drosera. But I must not run on about my hobby Horses.

Farewell | My dear Gray | Yours very sincerely | Charles Darwin

[1] Sterility and hybridisation are discussed in the conclusion of CD's paper, 'Dimorphic condition in *Primula*':

> The simple fact of two individuals of the same undoubted species, when homomorphically united, being as sterile as are many distinct species when crossed, will surprise those who look at sterility as a special endowment to keep created species distinct. Hybridizers have shown that individual plants of the same species vary in their sexual powers, so far that one individual will unite more readily than another individual of the same species with a distinct species. Seeing that we thus have a groundwork of variability in sexual power, and seeing that sterility of a peculiar kind has been acquired by the species of *Primula* to favour intercrossing, those who believe in the slow modification of specific forms will naturally ask themselves whether sterility may not have been slowly acquired for a distinct object, namely, to prevent two forms, whilst being fitted for distinct lines of life, becoming blended by marriage, and thus less well adapted for their new habits of life.

To Charles Lyell 23 [October 1861]

Down
23^{d}.

My dear Lyell

I ought to have returned your note by this morning's post, but I was too busy dissecting.— I suppose everyone takes your view that the water flowed out at head of valley when the lake existed. The "intermediate shelf" in Glen Roy, seemed to me on careful examination as plain as any shelf whatever; but I think I remember M^{r}. J[amieson] did not think so; it has been noticed by everyone who has visited the valley. There is no outlet known corresponding with this shelf; but as Mr Milne says there may be; & the whole valley ought to be searched for lateral outlets between the two upper shelves.— A man might spend his life there. I hope M[r]. J. will go there again; for it is an opprobrium to British Geologists, that it sh[d]. not be settled beyond dispute.

My difficulty is the sloping mass of matter, stratified & thick, at bottom of valley below lowest horizontal shelf: I think the river must have delivered detritus at infinitely many levels, by opening on a lake or arm of sea.— If there was terminal moraine at mouth of Spean to be *slowly* cut through all would be explained: I can hardly think ice would suffice. But if it were the sea, I cannot help a sneaking hope that the sea might have formed the horizontal shelves.—

Ever yours | C. Darwin

1862

[Thomas Henry Huxley delivered two lectures under the title 'Relation of man to the lower animals' at the Philosophical Institution of Edinburgh on 4 and 7 January 1862.]

To T. H. Huxley 14 [January 1862]

Down Bromley Kent
14th.

My dear Huxley

I am heartily glad of your success in the North, & thank you for your note & slip.— By Jove you have attacked Bigotry in its strong-hold. I thought you would have been mobbed. I am so glad that you will publish your Lectures. You seem to have kept a due medium between extreme boldness & caution.— I am heartily glad that all went off so well.—

I hope Mrs. Huxley is pretty well.— We have been a miserable family with 3 or 4 or 6 all in bed at the same time with virulent Influenza.— I have done nothing for nearly 3 weeks, & am much shaken.— . . .

I must say one word on Hybrid question.—[1] no doubt you are right that here is great hiatus in argument; yet I think you overrate it— you never allude to the excellent evidence of *varieties* of Verbascum & Nicotiana being partially sterile together. It is curious to me to read (as I have to day) the greatest crossing *Gardener* [Donald Beaton], utterly poop-poohing the distinction which *Botanists* make on this head, & insisting how frequently crossed *varieties* produce sterile offspring.— Do oblige me by reading latter half of my Primula paper in Lin. Journal for it leads me to suspect that sterility will hereafter have to be largely viewed as an acquired or *selected* character.— a view which I wish I had had facts to maintain in the Origin.— . . .

Farewell. I am poor weak wretch with trembling hands; so good night, & all good luck to you.— Ever yours | C. Darwin …

I find Brown-Sequard is largely with me, & will review in France the French Translation of the Origin.—

[1] See letter to T. H. Huxley, 11 January [1860].

From T. H. Huxley 20 January 1862

Jermyn St
Jany 20th 1862

My dear Darwin

The inclosed article which has been followed up by another more violent more scurrilously personal & more foolish, will prove to you that my labour has not been in vain—and that your views & mine are likely to be better ventilated in Scotland than they have been …[1]

I shall send a short reply to the 'Scotsman' for the purpose of further advertising the question—[2]

With regard to what are especially your doctrines—I spoke much more favourably than I am reputed to have done— I expressed no doubt as to their ultimate establishment—but as I particuly desire not to be misrepresented as an advocate trying to soften or explain away real dificulties—I did not in speaking enter in to the details of what is to be said in diminishing the weight of the hybrid difficulty— All this will be put fully when I print the Lecture—

The arguments put in your letter are those which I have urged to other people— of the opposite side—over & over again.

I have told my students that I entertain no doubt that twenty years experiments on pigeons conducted by a skilled physiologist instead of by a mere breeder—, would give us physiological species sterile *inter se* from a common stock—(& in this if I mistake not I go further than you do yourself) and I have told them that when these experiments have been performed I shall consider your views to have a complete physical basis—and to stand on as firm ground as any physiological theory whatever—…

I am constitutionally slow of adopting any theory that I must need stick by when I have once gone in for it—but for these two years I have been gravitating towards your doctrines & since the publication of your Primula paper with accelerated velocity— By about this time next year I expect to have shot past you— and to find you pitching into me for being more Darwinian than yourself— However, you have set me going & must just take the consequences, for I warn you I will stop at no point so long as clear reasoning will carry me further— …

Ever | Yours faithfully | T. H. Huxley …

[1] The *Witness*, the de facto organ of the evangelical Free Church of Scotland, carried an attack on Huxley's lectures on 14 January 1862.

[2] Huxley's letter appeared in the *Scotsman*, 24 January 1862, p. 2 (see *Correspondence* vol. 10, Appendix V).

To Asa Gray 22 January [1862]

Down Bromley Kent.—
Jan. 22^d.

My dear Gray

Your letter interested us much; for we are all curious to see how things look to you all, & a letter is something living.— But first thanks for your new cases of Dimorphism: new cases are tumbling in almost daily, but I shall never have time to work a quarter of them. You will have received before this my Primula paper, & will know the amount of evidence.— I have been ill with influenza (indeed we all have, for there have been 15 in bed in my household) & this has lost me 3 whole weeks, & delayed my little Orchid Book.— I fear that you expect in this opusculus much more than you will find— I look at it as a hobby-horse, which has given me great pleasure to ride. . . . I shall be very curious to hear what you think of it; for I have no idea whether it has been worth the trouble of getting up,—though the facts, I am sure, were worth *my own* while in making out—

… I forwarded your letter to Boott & to Hooker, from whom I had a long & capital letter this morning. He is working like a Horse. Here is a good joke, my book on Nat. Selection, he says, has made him an aristocrat in fact— he thinks breeding—the high breeding of the aristocracy—of the highest importance.—

Now for a few words on politics; but they shall be few, for we shall no longer agree, & alas & alas, I shall never receive another kind message from M^{rs}. Gray. I must own that the speeches & actions recently of your leading men (I regard little the newspapers), and especially the Boston Dinner have quite turned my stomach. I refer to Wilkes' being made a Hero for boarding an unarmed vessel.—to the Judges advice to him—& to your Governor triumphing at a shot being fired, right or wrong, across the bows of a British vessel.[1] It is well to make a clean breast of it at once; & I have begun to think whether it would not be well for the peace of the world, if you were split up into two or three nations. On the other hand I cannot bear the thought of the Slave-holders being triumphant; & it is really fearful to think of the difficulty of making a line of separation between the N. & the S., with armies, fortifications, & custom-houses without end with your retrograde tariff. Now I have done for myself in your eyes; & Mrs Gray will be indignant at having sent a kind message to so false a caitiff.—

Well I can't help my change of opinion— . . . Bad man, as you will think me, I shall always think of you with affection.— Here is an insult! I shall always think of you as an Englishman.

Ever yours very sincerely | Charles Darwin . . .

1 In November 1861, Charles Wilkes, captain of the Union vessel *San Jacinto*, had ordered the seizure of two Confederate envoys from the British mail packet, the *Trent*. A dinner in honour of Wilkes was held in Boston on 26 November, the details of which were reported in *The Times*, 10 December 1861. Several speeches were made praising Wilkes's action. The news that the affair had been settled and war avoided had reached England early in January.

From J. D. Hooker [31 January – 8 February 1862]

Dear Darwin

I wrote you a frightful screed the other day about the development of an Aristocracy being the necessary consequence of Natural Selection—& then burnt it—so you must take the will for the deed & be thankful! If ever we meet again we will talk it over—

I have a capital letter from Bates who is the only man I know that is "thinking out" your doctrines to any purpose— … I think I have driven Bates back to Nat. Selection as the only way of solving his difficulties.— I do not know when I have met a more interesting thing than his mimetic butterflies— I wish I had time to do the same thing with plants, which is quite feasable to a very considerable degree.

What the deuce can keep you so irritable about Owen: how I wish I could soothe you, I suppose it is the effect of your isolated life, & yet I dare say I am as insane upon some far less worthy score. My only care is to avoid owen— I can see that he *hates me* now with an intense *hate*— he fell foul of me at the Linnæan the other night in a most contemptible manner, & in so foolish a one that in half a dozen words of answer I set the whole society laughing at him. My God what an eye he fixed on me— Won't I catch it— of course I shall, but no worse than if I had not— what do I care …

Huxley has got into a most contemptible squabble with the Edinburgh newspapers, I really am astonished that he should notice such rubbish as they fulminate— the beauty of it is that no one in Edinburgh who reads either side sees the other & no one out of Edinburgh reads either! It is not like a Times controversy which every one reads

Ever Yrs | J D Hooker …

From Charles Kingsley 31 January 1862

Eversley Rectory, | Winchfield.

Private

My dear Mr. Darwin

I have just returned from Lord Ashburton's, where the Duke of Argyle, the B[isho]p of Oxford, & I, have naturally talked much about you & your book. As for the Bp. you know what he thinks— & more important, *you know what he* **knows**. The Duke is a very difft. mood; calm, liberal, & ready to hear all reason; though puzzled as every one must be, by a hundred new questions wh. you have opened.

What struck us on you & your theory, was, the shooting in the park of a pair of "blue Rocks", wh. I was called to decide on. There were several Men there who knew blue Rocks. The Duke said that the specimen was diff^{t} from the Blue Rock of the Hebrides—Young Baring[1] that it was difft from the B. R. of Gibraltar, & of his Norfolk Rabbit warrens (w^{h}. I don't believe from the specimens I have seen, to be a B. R. at all, but a stunted Stock dove, w^{h}. breeds in rabbit holes.), & I could hardly swear that this was a B. R. (as the keeper held) till I saw, but very weakly developed, the black bars on the wing coverts. . . .

My own view is—& I coolly stated it, fearless of consequences—that the specimen before me was only to be explained on your theory, & that Cushat, Stock doves & Blue Rock, had been once all one species—& I found—to shew how your views are steadily spreading—that of 5 or 6 men, only one regarded such a notion as absurd. If you want a specimen, I can get you one at once.

I want now to bore you on another matter. This great gulf between the quadrumana & man; & the absence of any record of species intermediate between man & the ape. It has come home to me with much force, that while *we* deny the existence of any such, the legends of most nations are full of them. Fauns, Satyrs, Inui, Elves, Dwarfs—we call them one minute mythological personages, the next conquered inferior races—& ignore the broad fact, that they are always represented as more bestial than man, & of violent sexual passion.

The mythology of every white race, as far as I know, contains these creatures, & I (who believe that every myth has an original nucleus of truth) think the fact very important.

The Inuus of the old Latins is obscure: but his name is from *inire*—sexual violence

The Faun of the Latins (or Romans, I dont know w^{h}.) has a monkey face, & hairy hind legs & body— the hind feet are traditionally those of a goat, the goat being the type of lust.

The Satyr of the Greeks is completely human, save an ape-face & a short tail—

The Elves Fairies & Dwarfs puzzle me, the 2 first being represented, originally, as of great beauty, the Elves *dark*, & the Fairies *fair*; & the Dwarfs as cunning magicians, & workers in metal—They may be really conquered aborigines.

The Hounuman, monkey God of India, & his monkey armies, who take part with the Brahminæ invaders, are now supposed to be a slave negro race, who joined the new Conquerors against their old masters. To me they point to some similar semi-human race. That such creatures sh^{d} have become divine, when they became rare, & a fetish worship paid to them—as happened in *all* the cases I have mentioned, is consonant with history—& is perhaps the only explanation of fetish-worship. The fear of a terrible, brutal, & mysterious creature, still lingering in the forests.

That they should have died out, by simple natural selection, before the superior white race, you & I can easily understand. . . .

I hope that you will not think me dreaming— To me, it seems strange that we are to deny that any Creatures intermediate between man & the ape ever existed, while our forefathers of every race, assure us that they did— . . .

At least, believe me | Ever, differing now, & now agreeing— | Yours most faithfully | C Kingsley

Eversley
January 31/62

[1] Possibly Alexander Hugh Baring, whose father lived at Buckenham Hall in Norfolk; Alexander was the nephew of William Bingham Baring, Lord Ashburton.

To Charles Kingsley 6 February [1862]

Down. | Bromley. | Kent. S.E.
Feb. 6th

My dear Mr Kingsley

I thank you sincerely for your letter.— I have been glad to hear about the Duke of Argyle, for ever since the Glasgow Brit. Assoc. when he was President, I have been his ardent admirer. What a fine thing it is to be a Duke: nobody but a Duke, the first time he geologised would have found a new formation; & the first time he botanised a new lichen to Britain.—

With respect to the pigeons, your remarks show me clearly (without seeing specimens, though I thank you for the kind offer) that the birds shot were the Stock Dove or C. Oenas, long confounded with the Cushat & Rock-pigeon. It is in some respects intermediate in appearance & habits; as it breeds in *holes* in trees & in rabbit-warrens. It is so far intermediate that it quite justifies what you say on all the forms being descendants of one.—

That is a grand & almost awful question on the genealogy of man to which you allude. It is not so awful & difficult to me, as it seems to be most, partly from familiarity & partly, I think, from having seen a good many Barbarians. I declare the thought, when I first saw in T. del Fuego a naked painted, shivering hideous savage, that my ancestors must have been somewhat similar beings, was at that time as revolting to me, nay more revolting than my present belief that an incomparably more remote ancestor was a hairy beast. Monkeys have downright good hearts, at least sometimes, as I could show, if I had space. I have long attended to this subject, & have materials for a curious essay on Human expression, & a little on the relation in mind of man to the lower animals. . . .

It is a very curious subject, that of the old myths; but you naturally with your classical & old-world knowledge lay more stress on such beliefs, than I do with all my profound ignorance. Very odd those accounts in India of the little hairy men! It is very true what you say about the higher races of men, when high enough, replacing & clearing off the lower races. In 500 years how the Anglo-saxon race will have spread & exterminated whole nations; & in consequence how much the Human race, viewed as a unit, will have risen in rank. Man is clearly an old-world, not an American, species; & if ever intermediate forms between him & unknown Quadrumana are found, I should expect they would be found in Tropical countries, probably islands. But what a chance if ever they are discovered: look at the French beds with the celts, & no fragment of a human bone.— It is indeed, as you say absurd to expect a history of the early stages of man in prehistoric times.—

I hope that I have not wearied you with my scribbling & with many thanks for your letter, I remain with much respect— | Yours sincerely | Charles Darwin . . .

[Heinrich Georg Bronn had translated *Origin* into German soon after its publication in November 1859; he added a preface of his own to the translation. This letter is translated from the original German.]

From Heinrich Georg Bronn [before 11 March 1862]

Most honoured Sir, . . .

I have read about a dozen reviews of your book in German, Dutch, English, and American journals, some favourable and many unfavourable. However, they have not altered *my own* opinion.

1) I see in your theory the only natural route to the final solution of the enigma of creation.

2) Yet, your theory contradicts the *current* state of our knowledge of the formation of *organic matter* from inorganic, of the *vitality* of organic matter and its organisation into organic *form* **without** *previous influence*

3). Our knowledge in this regard, however, is negative, and it *cannot be maintained* that future discoveries will not give us the kind of positive knowledge needed to be able to accept your theory. Even if your theory cannot be accepted at present, one cannot reject it *for all time*!

Among the many opinions of your **theory**, I have not found one that agrees with my own, but I consider mine correct. All judgments about your **book**, however, were favourable . . .

Please accept the assurance of my highest esteem. I have the honour to remain | your humble servant | H. G. Bronn

[On 15 May 1862, CD's book *On the various contrivances by which British and foreign orchids are fertilised by insects, and the good effects of intercrossing* (*Orchids*), was published.]

To Asa Gray 10–20 June [1862]

Down Bromley Kent
June 10th.

My dear Gray,

Your generous sympathy makes you overestimate what you have read of my orchid Book. But your letter of May 18th + 26th has given me an almost foolish amount of satisfaction. The subject interested me, I know, beyond its real value; but I had lately got to think that I had made myself a complete fool by publishing in a semi-popular form. Now I shall confidently defy the world. I have heard that Bentham & Oliver approve of it; but I have heard the opinion of no one else, whose opinion is worth a farthing. . . .

Enough & too much about my orchids, which are now again become beloved in my eyes, & which were quite lately accursed. . . .

Forgive me for one bit more trouble: I have a Boy with the collecting mania & it has taken the poor form of collecting Postage stamps: he is terribly eager for "Well, Fargo & Co Pony Express 2^d^ & 4^d^ stamp", & in a lesser degree "Blood's 1. Penny Envelope, 1, 3, & 10 cents". If you will make him this present you will give my dear little man as much pleasure, as a new & curious genus gives us old souls.[1]

Since this was written the above little man has been struck down with scarlet-fever; but thank God this morning the case has taken a mild form.—

I have just received your long notes on Cypripedium; you may believe how profoundly interesting they are to me. Will you not publish them, either in noticing my Book in Silliman['s Journal],[2] or otherwise? But your notes are more interesting than you will suppose, for since publishing I saw at Flower show, C. hirsutissimum, but could not touch it, but it seemed to me that the sterile anther entirely covered the passages by the anthers. I was amazed & saw clearly that there must be some quite distinct manner of fertilisation. But I did not think of insects crawling into flower; still less of different kind of pollen & in somewhat concave & viscid stigma. By Jove it is wonderful. You have hit on the same very idea which latterly has overpowered me, viz the exuberance of contrivances for same object: you will find this point discussed & attempted to be partly explained in the last Chapter. No doubt my volume contains much error: how curiously difficult it is, to be accurate, though I try my utmost. Your notes have been interested me beyond measure. I can now afford to d—d. my critics with ineffable complacency of mind. Cordial thanks for this benefit.—

It is surprising to me that you sh^d.^ have strength of mind to care for science, amidst the awful events daily occurring in your country. I daily look at the Times with almost as much interest as an American could do. When will peace come: it is dreadful to think of the desolation of large parts of your magnificent country; & all the speechless misery suffered by many. I hope & think it not unlikely that we English are wrong in concluding that it will take a long time for prosperity to return to you. It is an awful subject to reflect on.— Good Bye my dear friend.— …

I received 2 or 3 days ago a French Translation of the Origin by a Mad^elle.^ Royer,[3] who must be one of the cleverest & oddest women in Europe: is ardent Deist & hates Christianity, & declares that natural selection & the struggle for life will explain all morality, nature of man, politicks &c &c!!!. She makes some very curious & good hits, & says she shall publish a book on these subjects, & a strange production it will be. . . .

Yours cordially C. Darwin

[1] CD refers to his twelve-year-old son, Leonard Darwin.

[2] Charles Darwin, *De l'origine des espèces ou des lois du progrès chez les êtres organisées*, translated by C. A. Royer (Paris, 1862).

[3] The *American Journal of Science and Arts* was known as Silliman's journal after its editor, Benjamin Silliman.

[This letter is translated from the original French.]

From Alphonse de Candolle 13 June 1862

Geneva
13 June 1862.

My dear Sir

I wish to thank you very much for sending me your article *On the dimorphic condition of Primula.* It is very intriguing and rightly gives one cause for reflection, like everything you publish. I have instructed my son to prepare an extract for the Bibliothèque universelle (archives des sciences) and in fact he has edited it for one of the next issues.

The main fact, that of superior fertility by crossing of the two least similar forms, seems to me to be connected in physiology with only one known fact, that of the higher fertility and more vigorous offspring of individuals not related to each other, while in-breeding is unfavourable. This is quite mysterious—from a theoretical point of view, one would sooner have expected the opposite—but it is a fact.

I have not noted whether you sowed the seeds obtained from the different crossings of your Primula. One would like to know whether the two forms reappear in equal proportions from seeds resulting from different crosses, or whether a given kind of seed produces one form. You are right when you say that the same plant keeps its form of flowers from year to year, but what happens generation after generation? . . .

Will we soon have the great work that you announce as providing detailed evidence of facts mentioned in your book on the origin of species?[1] I await it with great impatience. In conclusion, after reading your book three or four times, sometimes entirely, sometimes partially, my view is close to Asa Gray's. I like your theory. It delights my mind. It is the only one that makes sense of very obscure questions, unapproachable by other paths—but we need proofs for it, especially regarding *natural selection.* The general hypothesis of indefinite transmission across centuries of forms with more or less marked modifications seems preferable to any other, but I am uncertain that natural selection is the means for it. There are so many factors that for a long time keep forms the same from generation to generation or that cause them to revert! It is so rare for a new form to be preserved without the protection of man! I know of no proven instance of the latter case. There are some, probably, but none has been proved, as far as I know. . . .

Please accept, dear Sir, the expression of my high regard and all my devotion. | Alph. de Candolle

1 Candolle refers to CD's explanation in *Origin*, p. 2, that the work was an abstract, without references or authorities. CD stated: 'No one can feel more sensible than I do of the necessity of hereafter publishing in detail all the facts, with references, on which my conclusions have been grounded; and I hope in a future work to do this.' The only part of the planned three-part work that was published during CD's lifetime was *Variation* (1868).

To Alphonse de Candolle 17 June [1862]

Down. | Bromley. | Kent. S.E.

June 17th

My dear Sir

I am extremely much obliged for your kind & very interesting letter. I am pleased that you are interested by the Primula case. Your questions & remarks show that you have gone to the root of the matter. I am now trying various analogous experiments on several plants & on the seedlings raised from the so-called heteromorphic & homomorphic unions; & the results (as far as I have yet seen; for the capsules are gathered, but not yet examined) are interesting; Whenever I publish I will do myself the pleasure of sending you a copy. . . .

You kindly enquire about my larger work; it does make progress, but very slowly owing to my own weak health & ill-health in my family. I have, also, been seduced to publish a small work on the Fertilisation of Orchids, which has taken up nearly ten months. As M[r] Bentham & Asa Gray think well of this Book, I have sent by this post a copy for you. One main object has been to show how wonderfully perfect the structure of plants is; another regards close breeding in & in, to which I see you have attended.— I am not at all surprised that you are not willing to admit natural selection: the subject hardly admits of direct proof or evidence. It will be believed in only by those who think that it connects & partly explains several large classes of facts: in the same way opticians admit the undulatory theory of light, though no one can prove the existence of ether or its undulations.—

. . . I am, also, rejoiced to hear that you have the intention of again returning to Geographical Distrib[n]. I believe few, or no one, can have read your truly great work with more care than I have;[1] & no one can feel more respect & admiration for it & its author.—

Pray believe me, my dear Sir | Yours sincerely & respectfully | Ch. Darwin

[1] Alphonse de Candolle, *Géographie botanique raisonnée ou exposition des faits principaux et des lois concernant la distribution géographique des plantes de l'époque actuelle* (Paris and Geneva, 1855).

To Asa Gray 23[–4] July [1862]

Down Bromley Kent
July 23[d]

My dear Gray

I received several days ago two large packets, but have as yet read only your letter; for we have been in fearful distress & I could attend to nothing. Our poor Boy had the rare case of second rash & sore throat, besides mischief in kidneys; & as if this was not enough a most serious attack of erysipelas with typhoid symptoms. I despaired of his life; but this evening he has eaten one mouthful

& I think has passed the crisis. He has lived on Port-wine every $\frac{3}{4}$ of an hour day & night. This evening to our astonishment he asked whether his stamps were safe & I told him of the one sent by you, & that he sh^{d}. see it tomorrow. He answered "I should awfully like to see it now"; so with difficulty he opened his eyelids & glanced at it & with a sigh of satisfaction said "all right".— Children are one's gretest happiness, but often & often a still greter misery. A man of science ought to have none,—perhaps not a wife; for then there would be nothing in this wide world worth caring for & a man might (whether he would is another question) work away like a Trojan.— ...

... I hear the French say that my paper on Primula is all pure imagination; but I cannot hear that this is grounded on any observations—

... Of all the carpenters for knocking the right nail on the head, you are the very best: no one else has perceived that my chief interest in my orchid book, has been that it was a "flank movement" on the enemy. . . .[1]

By the way one of my chief enemies (the sole one who has annoyed me) namely Owen, I hear has been lecturing on Birds, & admits that all have descended from *one*, & advances as his own idea that the oceanic wingless Birds have lost their wings by gradual disuse. He never alludes to me or only with bitter sneers ...

... I have managed to skim the news-paper, but had not heart to read all the bloody details. Good God what will the end be; perhaps we are too despondent here; but I must think you are too hopeful on your side of the water. I never believed the "canard" of the army of the Potomac having capitulated. My good dear wife & self are come to wish for Peace at any price.

Good Night my good friend. I will scribble no more— C. D.

One more word. I sh^{d} like to hear what you think about what I say in last Ch. of Orchid Book on the meaning & cause of the endless diversity of means for same general purpose.— It bears on design—that endless question—

Good Night Good Night. . . .

Farewell my good Friend | C. Darwin

[1] The intricate relationship between orchid structure and pollination by insects lent itself readily to arguments for design in nature.

To John Lubbock 5 September [1862]

Cliff Cottage | Bournemouth
Sept 5th

My dear Lubbock

Many thanks for your pleasant note in return for all my stupid trouble.— I did not fully appreciate your insect-diving-case before your last note;[1] nor had I any idea that the fact was new, though new to me. It is really very interesting. Of course you will publish an account of it. You will then say whether the insect can fly well through the air. My wife asked how did he find out that it stayed 4 hours under water without breathing; I answered at once "Mrs. Lubbock sat four hours watching". I wonder whether I am right— …

Ever dear Lubbock | Yours very sincerely | C. Darwin

See what it is to be well trained. Horace said to me yesterday, "if everyone would kill adders they would come to sting less". I answered "of course they would, for there would be fewer". He replied indignantly "I did not mean that; but the timid adders which run away would be saved, & in time they would never sting at all" Natural selection of cowards!

[1] Lubbock had told CD of his discovery of a swimming hymenopterous insect.

To Charles Lyell 14 October [1862]

Down. | Bromley. | Kent. S.E.
Oct 14th

My dear Lyell

I return Jamieson's capital letter. I have no comments, except to say that he has removed all my difficulties & that now & for ever more I give up & abominate Glen Roy & all its belongings.— It certainly is a splendid case & wonderful monument of the old Ice Period.— … How many have blundered over those horrid shelves! …

Ever yours | C. Darwin

P.S. I am rather overwhelmed with letters at present, & it has just occurred to me that perhaps you will forward my note to Mr Jamieson; as it will show that I entirely yield. I do believe every word in my Glen Roy paper is false—

To Asa Gray 16 October [1862]

Down Bromley Kent
Oct. 16th

My dear Gray …

Now that we are at home again, I have begun dull steady work on "Variation under Domestication"; but alas & alas pottering over plants is much better sport.—

By the way at Bournemouth, for the want of something else to do, I worked a bit at my old friend Drosera: I took to testing all sorts of fluids, which are not corrosive & do not, I believe, act on ordinary organic compounds, but do act on the nervous system of animals; & I declare I am coming to the conclusion that plants or at least Drosera, must have something closely analogous to nervous matter. It was pretty to see effect of acetate of strychnine, how it stopped all movement; & how acetate of morphia greatly dulled & retarted movement. I think I shall some day pursue this subject.—

Another little point has interested me, viz finding such a number of natural hybrids between two species of Verbascum; & linking V. thapsus & lychnitis closely together. They are all utterly sterile. This fact has given Hooker, to whom I told it, a fit of the horrors.— …

Many thanks for sending the article in Daily News, which we read aloud in Family conclave. Our verdict was, that the N. was fully justified in going to war with the S.; but that as soon as it was plain that there was no majority in the S. for ReUnion, you ought, after your victories in Kentucky & Tennessee, to have made peace & agreed to a divorce. How curious it is that you all seem to believe that you can annex the South; whilst on this side of the Atlantic, it is the almost universal opinion that this is utterly impossible. If I could believe that your Presidents proclamation would have any effect, it would make a great alteration in my wishes; I would then run the risk of your seizing on Canada (I wish with all my heart it was an independent country) & declaring war against us. But slavery seems to me to grow a more hopeless curse. How detestably the special correspondent of the Times writes on the subject; the man has not a shade of feeling against slavery. This war of yours, however it may end, is a fearful evil to the whole world; & its evil effect will, I must think, be felt for years.—

I can see already it has produced wide spread feeling in favour of aristocracy & Monarchism: no one in England will speak for years in favour of the people governing themselves.

Well good night.— Do not be indignant with me & do not let Mrs Gray be more indignant than she can help.— Good Night & farewell | Yours cordially | Ch. Darwin . . .

[Writing in 1869 in 'Fertilization of orchids', p. 153, CD said of the genus *Acropera*:

> I have committed a great error about this genus, in supposing that the sexes were separate. Mr. J. Scott, of the Royal Botanic Garden of Edinburgh, soon convinced me that it was a hermaphrodite, by sending me capsules containing good seed, which he had obtained by fertilizing some flowers with pollen from the same plant.]

To John Scott 12 November [1862]

Down. | Bromley. | Kent. S.E.
Nov. 12th

Dear Sir

I thank you most sincerely for your kindness in writing to me, & for very interesting letter. Your fact has surprised me greatly, & has alarmed me not a little, for if I am in error about Acropera I may be in error about Catasetum. Yet when I call to mind the state of the placentæ in A. luteola, I am astonished that they should produce ovules. You will see in my book that I state that I did not look at the ovarium of A. Loddegesii. Would you have the kindness to send me word, which end of the ovarium is meant by *apex* (that nearest the flower?) for I must try & get this species from Kew & look at its ovarium.

I shall be extremely curious to hear whether the fruit, which is now maturing, produces a large number of *good* & plump seed; perhaps you may have seen the ripe capsules of other Vandeæ & may be able to form some conjecture what it ought to produce. In the young unfertilised ovaria of many Vandeæ, there seemed an infinitude of ovules. In desperation it occurs to me as just possible, as almost every thing in nature goes by gradation, that a properly male flower might occasionally produce a few seeds, in same manner as female plants sometimes produce a little pollen.

All your remarks seem to me excellent & very interesting & I again thank you for your kindness in writing to me.

I am pleased to observe that my description of the structure of Acropera seems to agree pretty well with what you have observed. Does it not strike you as very difficult to understand how insects remove the pollinia & carry them to the stigmas?— Your suggestion that the mouth of the stigmatic cavity may become charged with viscid matter & thus secure the pollinia, & that the pollen-tubes may then protrude, seems very ingenious & new to me; but it would be very anomalous in Orchids, i.e. as far as I have seen. No doubt, however, though I tried my best, that I shall be proved wrong in many points. Botany is a new subject to me.—

With respect to the protrusion of pollen-tubes, you might like to hear (if you do not already know the fact) that, as I saw this summer, in the little imperfect flowers of Viola & Oxalis, which never open, that the pollen-tubes always come out of the pollen-grain, *whilst still in the anthers*, & direct themselves in a beautiful manner to the stigma seated at some little distance.—

I hope that you will continue your very interesting observations, & I beg leave to remain | Dear Sir | Yours faithfully & obliged | Charles Darwin

[Henry Walter Bates applied the theory of natural selection to explain how some species survived by mimicking species that were unpalatable to predators. The paper CD discusses in his letter is 'Contributions to an insect fauna of the Amazon valley. *Lepidoptera: Heliconidæ*', *Transactions of the Linnean Society of London* n.s. 5 (1862): 495–566. CD published a review of this paper ('Review of Bates on mimetic butterflies').]

To H. W. Bates   20 November [1862]

Down Bromley Kent
Nov. 20th

Dear Bates

I have just finished after several reads your Paper. In my opinion it is one of the most remarkable & admirable papers I ever read in my life. The mimetic cases are truly marvellous & you connect excellently a host of analogous facts. The illustrations are beautiful & seem very well chosen; but it would have saved the reader not a little trouble, if the name of each had been engraved below each separate figure; no doubt this would have put the engraver into fits, as it would have destroyed beauty of Plate. I am not at all surprised at such a paper having consumed much time. I rejoice that I passed over whole subject in the Origin, for I sh[d]. have made a precious mess of it. You have most clearly stated & solved a wonderful problem.—

No doubt with most people this will be the cream of the paper; but I am not sure that all your facts & reasoning on variation & on the segregation of complete & semi-complete species is not really more, or at least as valuable a part.— I never conceived the process nearly so clearly before; one feels present at the creation of new forms.— I wish, however, you had enlarged a little more on the pairing of similar varieties; a rather more numerous body of facts seems here wanted.

Then again what a host of curious miscellaneous observations there are,—as on related sexual & individual variability you give; these will some day, if I live, be a treasure to me.—

With respect to mimetic resemblance being so common with insects; do you not think it may be connected with their small size; they cannot defend themselves;— they cannot escape by flight at least from Birds; therefore they escape by trickery & deception?

I have one serious criticism to make & that is about title of paper; I cannot but think that you ought to have called prominent attention in it to the mimetic resemblances.— Your paper is too good to be largely appreciated by the mob of naturalists without souls; but rely on it, that it will have *lasting* value, & I cordially congratulate you on your first great work. You will find, I sh[d]. think, that Wallace will fully appreciate it.— . . .

Believe me Dear Bates | Yours very sincerely | Ch. Darwin . . .

From J. D. Hooker 26 November 1862

Royal Gardens Kew
Nov 26/62

D[r]. Darwin

I return A Grays letter with a thousand thanks . . .

. . . his whole letter breathes an accursed spirit of jealousy of our strength, & Americas weakness.

The more I reflect, the more sure I am that America will never settle untill she has the equivalent of an Aristocracy (used in best sense) wherefrom to chuse able Governors & statesmen. There is no more certain fruits of your doctrines than this—that the laws of nature lead infallibly to an aristocracy, as the only security for a *settled condition of improvement*— What has prevented America having one of same sort hitherto?, but the incessant pouring in of democratic* elements from the West—which has prevented the sorting of the masses, & frustrated all good effects of Natural Selection. * By a democracy in bad sense I mean a tendency to reduce the better to the worse level

By the way when you have any difficulties such as believing too much in action of physical conditions, you must do as the parsons tell their flocks—come to **me** or some other wise & discreet &c &c &c— . . .

I am still very strong in holding to impotence of crossing with respect to *origin* of species— I regard variation as so illimitable in [animals]— You must remember that it is neither crossing nor N. Selection that has **made** so many divergent human individuals, but simply **variation**: Nat. Sel. no doubt has hastened the process, intensified it so to speak, has regulated the lines places &c &c &c. in which & to which the races have run & led, & the number of each & so forth;—but, given a pair of individuals with power to propagate, & infinite span to procreate in, so that not one be lost,—or that in short Nat. Sel. is not called on to play a part at all & I maintain that after n generations you will have extreme individuals as totally unlike one another as if Nat Sel. had extinguished half— If once you hold that Nat. Sel. can *make* a difference, ie *create* a character, your whole doctrine tumbles to the ground— N.S. is as powerless as physical causes to make a variation;—the law that "like shall *not* produce like" is at the bottom of all, & is as inscrutable as life itself. This it is that Lyell & I feel you have failed to convey with force enough to us & the public: & this is at the bottom of half the infidelity of the scientific world to your doctrine. You have not, as you ought, begun by attacking old false doctrines, that "like does produce like" the first chapter of your book should have been devoted to this & to nothing else. But there is some truth I now see in the objection to you, that you make N.S. the "Deus ex machina." for you do somehow seem to do it.—by neglecting to dwell on the *facts* of infinite incessant variation,— Your 8 children are really all **totally** unlike one another they agree **Exactly in no one property** how is this? you answer that they display the inherited differences of different progenitors—well—but go back, & back & back in time & you are driven at last to your original pair for origin of differences, & logically you must grant, that the differences between the original ♂ & ♀ of your species were = the su⟨m⟩ of the extreme differences between the most dissimilar existing individuals of your species!—or that the latter varied from some inherent law that had them. Now am not I a cool fish to lecture you so glibly— . . .

. . . your | dear friend
J D Hooker . . .

To J. D. Hooker [after 26] November [1862]

Down
Nov. 20th[1]

My dear Hooker

Your last letter has interested me to an extraordinary degree, & your truly parsonic advice "some other wise & discreet person" &c, amused us not a little.— I will put a "concrete" case to show what I think A. Gray believes about crossing & what I believe.— If 1000 pigeons were bred together in cage for ten 1000 years, their number not being allowed to increase by *chance* killing, then from mutual intercrossing no varieties would arise; but if each pigeon were a self-fertilising hermaphrodite a multitude of varieties would arise. This I believe is common effect of crossing, viz the obliteration of incipient varieties. I do not deny that when two marked varieties have been produced; their crossing will produce a third or more *intermediate* varieties. Possibly or probably with domestic varieties, with strong tendency to vary, the act of crossing tends to give rise to *new* characters; & thus a third or more races, not strictly intermediate, may be produced. But there is heavy evidence against *new* characters arising from crossing wild forms; only intermediate races are then produced.— Now do you agree thus far? if not, it is no use arguing, we must come to swearing, & I am convinced I can swear harder than you. ∴ I am right. Q.E.D.—

If the number of 1000 pigeons were prevented increasing, not by chance killing, but by, say, all the shorter-beaked birds being killed, then the **whole** body would come to have longer beaks. Do you agree?

Thirdly, if 1000 pigeons were kept in hot country, & another 1000 in cold country, & fed on different food & confined in different size aviary & kept constant in number by *chance* killing, then I sh^d. expect as rather probable that after ten 1000 years, the two bodies would differ slighty in size, colour & perhaps other trifling characters. This I sh^d. call the *direct* action of physical conditions. By this action I wish to imply that the innate vital forces are somehow led to act rather differently in the two cases. Just as heat will allow or cause two elements to combine, which otherwise would not have combined.— I sh^d. be especially obliged if you would tell me what you think on this head.—

But the part of your letter which fairly pitched me head over heels with astonishment; is that where you state that every single difference which we see might have occurred without any selection. I do & have always fully agreed; but you have got right round the subject & viewed it from an entirely opposite & new side, & when you took me there, I was astounded. When I say I agree, I must make proviso, that under your view, as now, each form long remains adapted to certain fixed conditions & that the conditions of life are in long run changeable; & 2^d, which is more important that each individual form is a self-fertilising hermaphrodite, so that each hair-breadth variation is not lost by intercrossing. Your manner of putting case would be even more striking than it is, if the mind could grapple with such numbers— it is grappling with eternity— think of each of a thousand seeds bringing forth its plant, & then each a thousand. A globe stretching to furthest fixed star would very soon be covered. I cannot even grapple with idea even with races of dogs, cattle, pigeons or fowls; & here all must admit & see the accurate strictness of your illustration.—

Such men, as you & Lyell thinking that I make too much of a Deus of N. Selection is conclusive against me.— Yet I hardly know how I could have put in, in all parts of my Book, stronger sentences. The title, as you once pointed out, might have been better. No one ever objects to agriculturalists using the strongest language about their selection; yet every breeder knows that he does not produce the modification which he selects. My enormous difficulty for years was to understand adaptation, & this made me, I cannot but think rightly, insist so much on N. Selection. God forgive me for writing at such length; but you cannot tell how much your letter has interested me, & how important it is for me with my present Book in hand to try & get clear ideas. Do think a bit about what is meant by direct action of physical conditions. I do not mean *whether* they act; my facts will throw some light on this. I am collecting all cases of "*bud*-variations in contradistinction to 'seed-variation'" (do you like this term for what *some* gardeners call "sports"): these eliminate all effect of crossing.— Pray remember how much I value your opinion, as the clearest & most original I ever get.— . . .

Ever yours | My dear Hooker | C. Darwin

[1] Although CD wrote 'Nov. 20th', this letter is evidently a reply to the letter from J. D. Hooker, 26 November 1862.

To W. B. Tegetmeier 27 [December 1862]

Down Bromley Kent
27th

My dear Sir

I . . . am heartily glad to hear of R.S. making so good a move.[1] I am, however, not sanguine of success.— The present plan is to try whether any existing breeds happen to have acquired accidentally any degree of sterility;—but to this point hereafter. . . .

I will suggest an . . . experiment, which I have had for two years in my Experimental book with "be sure & try". but which as my health gets yearly weaker & weaker & my other work increases, I suppose I shall never try. Permit me to add that if 5£ would cover expences of experiment, I shd. be delighted to give it & you could publish result *if there be any result*. I crossed Spanish Cock (your bird) & white Silk hen & got plenty of eggs & chickens; but two of these *seemed* to be quite sterile.

I was then sadly overdone with work but have ever since much reproached myself, that I did not preserve & carefully test the procretive power of these hens.— Now if you are inclined to get a Spanish Cock & a couple of *white* Silk hens, I shall be most grateful to hear whether the offspring breed well; they will prove, I think, not hardy; if they sh[d] prove sterile,, which I can hardly believe, they will anyhow do for the pot.—

If you do try this; how would it be to put a silk cock to your curious silky Cochin Hen; so as to get a *big* Silk breed; it would be curious if you could get silky fowl with bright colours— I believe a silk **hen** crossed by any other breed never give silky feather. A cross from Silk Cock & Cochin Silk Hen ought to give silky feathers & probably bright colours.—

I have been led lately from experiments (not published) on Dimorphism to reflect much on sterility from Hybridism & *partially* to change the opinion given in Origin. I have now letters out enquiring on following point, implied in the experiment, which seems to me well-worth trying, but too laborious ever to be attempted. I would ask every Pigeon & Fowl Fancier, whether they have ever observed *in the same breed*, a cock A paired to a hen B, which did not produce young. Then I would get cock A & match it to a hen of its nearest blood; & hen B to its nearest blood. I would then match the offspring of A (viz a, b, c, d, e) to the offspring of B, (viz f, g, h, i, j)—& all these children which were fertile together should be destroyed until I found, one, (say *a*) which was not quite fertile with (say *i*). Then *a* & *i* sh[d] be preserved & paired with their parents A & B, so as to try & get two families, which would not unite together; but the members **within** each family being fertile together. This would probably be quite hopeless; but he who could effect this, would, I believe, solve the problem of Sterility from Hybridism.—

If you sh[d] ever hear of *individual* fowls or pigeons which are sterile together, I sh[d] be very grateful to hear of case. It is parallel case to those recorded of a man *not* impotent long living with a woman who remained childless; the husband died & the woman married again & had plenty of children. *Apparently* (by no means certainly) this first man & woman were dissimilar in their sexual organisation. I conceive it possible that their offspring (if both had married again & both had children[)] would be sexually dissimilar like their parents or sterile together.—

Pray forgive my dreadful writing; I have been very unwell all day, & have no strength to rewrite this scrawl.— I am working slowly on, & I suppose in 3 or 4 months shall be ready for M.S. of Fowls.

My dear Sir | Yours very sincerely | C. Darwin
I am sure I do not know whether any human being could understand or read this shameful scrawl.—

[1] On 1 December 1862, the council of the Royal Society of London resolved to grant £10 to Tegetmeier for 'experiments on the cross-breeding of pigeons' (Royal Society, Council minutes, 1 December 1862).

To T. H. Huxley 28 December [1862]

Down Bromley Kent
Dec. 28th

My dear Huxley
... All that I said about the little book[1] is strictly my opinion; it is in every way excellent & cannot fail to do good the wider it is circulated. Whether it is worth your while to give up time to it, is another question for you alone to decide; that it will do good for the subject is beyond all question. I do not think a dunce exists, who could not understand it; & that is a bold saying after the extent to which I have been misunderstood.

I did not understand what you required about sterility: assuredly facts given do not go nearly so far. We differ so much that it is no use arguing. To get the degree of sterility you expect in recently formed varieties seems to me simply hopeless. It seems to me almost like those naturalists who declare they will never believe that one species turns into another till they see every stage in process.—

I have heard from Tegetmeier & have given him the results of my crosses of the birds which he proposes to try, & have told him how alone I think the experiment could be tried with faintest hope of success. Namely to get if possible case of two birds which when paired were unproductive, yet neither impotent; for instance I had this morning I had a letter with case of Hereford Heifer which seemed to be after repeated trials sterile with one particular far from impotent Bull, but not with another Bull..— But it is too long a story— it is to attempt to make two strains, both fertile, & yet sterile when one of one strain is crossed with one of the other strain. But the difficulty & distinction of the fertile individuals would be beyond calculation.—

As far as I see Tegetmeiers plan would simply test whether two existing breeds are now in any slight degree sterile; which has already been largely tested: not that I dispute good of retesting. . . .

I am tired—so good night | Ever yours very truly | C. Darwin

[1] Thomas Henry Huxley, *On our knowledge of the causes of the phenomena of organic nature* (London, 1862).

1863

[In 1861, a fossil was discovered in the Jurassic lithographic limestone near Solenhofen (now Solnhofen), Bavaria, that displayed both reptilian and bird-like features, including feathers connected to a long, jointed tail. It was named *Archaeopteryx macrura* by Richard Owen. The fossil achieved considerable notoriety as a 'feathered reptile', was purchased by Owen on behalf of the British Museum for the sum of £400, and was brought to London in October 1862.]

From Hugh Falconer 3 January [1863]

21 Park Crescent N.W.
3d. Jany.

My Dear Darwin

A happy new year to you—and many happy returns of the season to you and yours! . . .

I was sorry to hear from your brother, of the efflorescence which has been troubling you—and which he tells me is one of the reasons, that has prevented you from coming to town. You were never more missed—at any rate by me—for there has been this grand *Darwinian* case of the *Archæopteryx* for you and me to have a long jaw about. Had the Solenhofen quarries been commissioned—by august command—to turn out a strange being à la Darwin—it could not have executed the behest more handsomely—than in the *Archæopteryx*. This is sober earnest—and that you should not have been in to town—and see it and talk over it with me, is a criminal proceeding. You are not to put your faith in the slip-shod and hasty account of it given to the Royal Society.[1] It is a much more astounding creature—than has entered into the the conception of the describer—who compares it with the Raptores & Passeres. & Gallinaceæ, as a round winged (like the last) 'Bird of flight.' It actually had at least two long free digits to the fore limb—and those digits bearing claws as long and strong as those on the hind leg. Couple this with the long tail—and other odd things,—which I reserve for a jaw—and you will have the sort of misbegotten-bird-creature—the dawn of an oncoming conception à la Darwin. But I will not say more about it till you show yourself in town. . . .

Lyell is hard at work in finishing his book.[2] I fancy he has had much to alter and adjust—and I have got a quagmire-*ish* kind of feeling, that he—and the Subject Homo—will be bogged in it—alike.

My Dear Darwin | Yours very Sin[ly] | H Falconer

P.S. The last parag. is entre nous—so please do not repeat it.—

1 Richard Owen, 'On the *Archaeopteryx* of von Meyer, with a description of the fossil remains of a long-tailed species, from the lithographic stone of Solenhofen', *Philosophical Transactions of the Royal Society of London* 153 (1863): 33–47.

2 Charles Lyell, *The geological evidences of the antiquity of man with remarks on theories of the origin of species by variation* (London, 1863).

From John Scott 6 January 1863

Botanic Gardens, Edinburgh
January 6th 1863.

Sir,

I send off by train to-night a small box with three plants of P. Scotica and three of P. farinosa. I am sorry that I cannot get any more for you at present, but if I can possibly do so in the course of the ensuing spring; you may depend upon receiving them! On one of the plants of P. Scotica, you will observe a few capsules. Perhaps they may not contain many perfect seeds being the produce of late autumn flowers, in cold, damp weather. . . .

Had I remembered the *Begonia frigida*, I should have asked M[r]. M[c]Nab, before he visited Kew last to have seen if they could have spared a plant for the garden here, but now it is too late for this season. I am sorry as it would be an interesting subject for experiment, and I really feel shy in my circumstances to ask M[r]. M[c]Nab specially to see whether they could afford us a plant of it or not. And here I may take the opportunity of stating in reply to your kind enquiries into my personal circumstances, that in consequence of certain loses in business sustained by my parents, we were thus brought down from easy circumstances, to a comparatively humble position. My parents died while I was very young, leaving me in charge of a relative. After receiving an ordinary education, I became a gardener—more for the purpose of gratifying a predilection for Natural History—than any love for this line of life. But I thus thought that I might have better opportunities for pursuing these branches of Science, than by any other I had it in my power to engage in. And, so far, I am happy to say—I have not been disappointed, and I now through the kindness of Prof. Balfour, and Mr. McNab, enjoy great facilities for such pursuits. At present, I have the charge of the propagating department in the Botanic Gardens here, and I have thus excellent opportunities for performing experiments, having a sufficient range of temperature for all ordinary purposes. I may state that I was all but engaged a few months back by Prof. Balfour, to go to Madras, as Superintendent of the Horticultural Gardens there. I was ultimately disappointed of this situation, however, in consequence of the re-engagement of the then Superintendent. It is therefore hard to say what part of the world I may next be sent to, but time will show; And then, perhaps you will kindly favour me with a few suggestions hints, and advices, on observations that I may then have it in my power to make. And now by giving you excuse for this detailed account of personal affairs I will only ask you in conclusion to excuse my non-reference to one or two points in your last, which I will do at an early opportunity.

I remain | Sir, | Yours very respectfully | John Scott. . . .

To T. H. Huxley 10 [January 1863]

Down Bromley Kent
Dec. 10th

My dear Huxley

You will be weary of notes from me about the little book of yours. It is lucky for me that I expressed, before reading no VI, my opinion of its absolute excellence, & of its being well worth wide distribution & worth correction (not that I see where you could improve), if you thought it worth your valuable time. Had I read no VI, even a rudiment of modesty would, or ought to, have stopped me saying so much. Though I have been well abused, yet I have had so much praise, that I have become a gourmand, both as to capacity & taste; & I really did not think that mortal man could have tickled my palate in the exquisite manner with which you have done the job. So I am an old ass, & nothing more need be said about this.— I agree entirely with all your reservations about accepting the doctrine, & you might have gone further with further safety & truth. Of course I do not **wholly** agree about sterility. I hate beyon[d] all things finding myself in disagreement with any capable judge, when the premises are the Same; & yet this will occasionally happen.

Thinking over my former letter to you, I fancied (but I now doubt) that I had partly found out cause of our disagreement, & I attributed it to your naturally thinking most about animals, with which the sterility of the hybrids is much more conspicuous than the lessened fertility of the first cross. Indeed this could hardly be ascertained with mammals, except by comparing the product of whole life; & as far as I know this has only been ascertained in case of Horse & ass, which do produce fewer offspring in lifetime than in pure breeding. In plants the test of first cross seems as fair, as test of sterility of hybrids. And this latter test applies, I will maintain to the death, to the crossing of vars. of Verbascum & vars, selected vars, of Zea.— You will say go to the Devil & hold your tongue.— No I will not hold my tongue; for I must add that after going for my present book all through domestic animals; I have come to conclusion that there are almost certainly several cases of 2 or 3 or more species blended together & now perfectly fertile together. Hence I conclude that there must be something in domestication,—perhaps the less stable conditions,—the very cause which induces so much variability,—, which eliminates the natural sterility of species, when crossed. If so, we can see how unlikely that sterility should arise between domestic races. Now I will hold my tongue.— . . .

. . . Do not waste your time in answering this.—

Ever yours | C. Darwin

From James Dwight Dana 5 February 1863

New Haven,
Feby 5. 1863.

My dear Mr Darwin— . . .

I hope that ere this you have the copy of the Geology[1] . . . I have still to report your book[2] unread; for my head has all it can now do in my college duties.

I have thought that I ought to state to you the ground for my assertion on page 602, that Geology has not afforded facts that sustain the view that the system of life has been evolved through a method of development from species to Species.— There are three difficulties that weigh on my mind, and I will mention them.

1. The absence, in the great majority of cases, of those transitions by small differences required by such a theory.— As the life of America & Europe has been with few exceptions independent, one of the other, it is right to look for the transitions on each Continent separately.— The reply to this difficulty is that the Science of Geology is comparatively new and facts are daily multiplying. But this admits the proposition that Geology does not yet afford the facts required.—

2. The fact of the commencement of types in some cases by their higher groups of species instead of the lower.— As fishes began with the Selachians or Sharks (the highest order of fishes[)] & the Ganoids, which are above the true level of the fish between fishes & Reptiles. In the introduction of land plants, there were Acrogens & Conifers and intermediate types, but not the lower grade of mosses—seemingly the natural stepping-stone from the Seaweeds. The fishes, Lepidodendron, Sigillarids, are examples of those intermediate or *comprehensive* types, with which great groups often began, and seem to explain the true relations of such types: —that they were *not transitional* forms in the system of life, but rather the *commencing* forms of a type.— If I advocated your theory, I think I should take the ground that there were certain original points of divergence from time to time introduced into the System, as indicated by the Comprehensive types.

3. The fact that with the transitions in the strata & formations, the exterminations of species often cut the threads of genera, families & tribes, and sometimes, also, of the higher groups of Orders, classes and even Subkingdoms; and yet the threads have been started again in new species. The transition after the Carboniferous age was one apparently of complete extermination both in America & Europe, where all threads were cut; & yet life was reinstated and partly by renewing with species old genera in all the classes & subkingdoms, besides adding new types.

You thus see that I have not spoken positively on page 602, without thinking I had some foundation for it. I speak merely of the geological facts that bear on the or any theory of development, not of facts from other Sources.— . . .

With earnest wishes for your health and happiness, I remain | Sincerely yours | James D. Dana . . .

[1] James Dwight Dana, *Manual of geology: treating of the principles of the science with special reference to American geological history, for the use of colleges, academies, and schools of science* (Philadelphia and London, 1863).
[2] CD had sent Dana a presentation copy of *Origin* in November 1859.

From W. B. Tegetmeier 18 February 1863

Muswell Hill | N.
Feby 18/63

My dear Sir

I hoped to have the pleasure of seeing you at the Linnean society, and much regret the cause of your absence

There were one or two points connected with my little experiment that I should have been very glad to have asked you respecting, such as your suggestion to employ Carriers and Turbits, instead of Barbs and Fantails.—

In accordance with your suggestion I have procured some Silk fowls (two very good black skinned hens) and a Spanish Cock. I am devoting a small paddock to them and will hatch several clutches of chickens— Do you think that the two hens will be sufficient or can you suggest any modification of the experiment.— Would it be desirable or necessary to have a second set to cross their offspring with those of the first— If so I would make arrangements to procure them— —

Trusting to hear of improvement in your health | Believe me | Very truly yours | W B Tegetmeier

To W. B. Tegetmeier 19 February [1863]

Down Bromley Kent
Feb. 19th.

My dear Sir

I am delighted to hear that you have the Fowls: as soon as you have chickens you could kill off the old Birds. I shd. think the

3 ample.— It would be better to cross some cocks & Hens of the half-breds from the two nests; so as not to cross *full* brother & sister. I have not much hope that they will be partly or wholly sterile, yet after what happened to me, I sh^{d}. never have been easy without a trial.—[1]

I suggested Turbits, because statements have been published that they are sometimes sterile with other breeds, & I mentioned Carriers, merely as a very distinct breed. I thought Barbs & Fantails bad solely because I had made several crosses & found the $\frac{1}{2}$ breds perfectly fertile,—even brother & sister together. Did I send you (*I cannot remember*) a M.S. list of crosses; if so for Heaven sake return it.— I get *slowly* on with my work; but am never idle.—

I much wish I could have seen you at Linn. Soc; but I was that day very unwell.— Pray do not forget to ask Poultry & Pigeon men (especially latter) whether they have ever matched two birds (for instance two almonds, Tumblers) & could not get them to breed, but afterwards found that both birds would breed when otherwise matched.—

I hope the world goes pretty well with you.—

My dear Sir | Yours sincerely | C. Darwin

[1] CD had previously conducted the cross that Tegetmeier was about to undertake, between a Spanish cock and a white silk hen. See letter to W. B. Tegetmeier, 27 [December 1862].

To J. D. Dana 20 February [1863]

Down. | Bromley. | Kent. S.E.
Feb. 20th

My dear Mr Dana

I received a few days ago your Book & this morning your pamphlet on Man[1] & your kind letter. I am heartily sorry that your head is not yet strong, & whatever you do, do not again overwork yourself. Your book is a monument of labour, though I have as yet only just turned over the pages. . . .

With respect to the change of species I fully admit your objections are perfectly valid. I have noticed them, excepting one of separation of countries, on which perhaps we differ a *little.* I admit, that if we really now know the beginning of life on this planet, it is absolutely fatal to my views; I admit the same, if the geological record is not *excessively* imperfect; & I further admit that the a priori probability is that no being lived below our Cambrian era. Nevertheless I grow yearly more convinced of the general (with much incidental error) truth of my views: I believe in this from finding that my views embrace so many phenomena & explain them to a large extent. I am continually pleased by hearing of naturalists (within the last month I have heard of four) who have come round to a large extent to the belief of the modification— of Species.— As my book has been lately somewhat attended to, perhaps it would have been better, if when you condemn all such views, you had stated that you had not been able yet to read it. But pray do not suppose that I think for one instant that with your strong & slowly-acquired convictions, & immense knowledge, you would have been converted. The utmost that I could have hoped would have been that you might possibly have been here or there staggered. Indeed I should not much value any sudden conversion; for I remember well how many long years I fought against my present belief.—

With respect to Dr. Falconer, I fear I ought not to have said anything, as he lately told me that he shd. not interfere till Prof. Owen had published; so please do not repeat what I said. I daresay Owen will work out everything carefully before he publishes in detail; but there is little doubt he was at *first* very careless. He overlooked a jaw with teeth which however may possibly not have belonged to this marvellous Bird, with its long tail & fingers to its wings. As Birds are so isolated this case, as you may suppose, has pleased me.— You will see in Lyell's book that Owen has made a fearful mistake (not discovered by himself) about the British Eocene monkey.— He has made such mistakes about the Elephants & Rhinoceroses, that I declare I am getting fearful of trusting him. He has done the work of a giant; but I fear he has been too ambitious & not given time enough to most of his work.— I have not yet read Huxley's book;[2] but I hear it is very striking; but you will highly disapprove of it.—

With every good wish | pray believe me | Yours very sincerely | Ch. Darwin …

[1] J. D. Dana, *Manual of geology* (Philadelphia, 1863); 'On the higher subdivisions in the classification of mammals', *American Journal of Science and Arts* 2d ser. 35 (1863): 65–71.

[2] Thomas Henry Huxley, *Evidence as to man's place in nature* (London, 1863). CD also refers to Charles Lyell, *Geological evidences of the antiquity of man* (London, 1863).

To Charles Lyell 6 March [1863]

Down
March 6th

My dear Lyell.

Putting you off was a great & bitter disappointment. . . . But I had no choice, & I much fear that Emma is right & that I must knock off all work & all of us go to Malvern for two months. I get on with nothing.— . . .

I have been of course, deeply interested by your Book:[1] I have hardly any remarks worth sending, but will scribble a little on what most interested me. But I will first get out what I hate saying, viz that I have been greatly disappointed that you have not given judgment & spoken fairly out what you think about the derivation of Species. I shd. have been contented if you had boldly said that species have not been separately created, & had thrown as much doubt as you like on how far variation & N. Selection suffices. . . . I think the Parthenon is right that you will leave the Public in a fog. No doubt they may infer that as you give more space to myself, Wallace & Hooker than to Lamarck, you think more of us. But I had always thought that your judgment would have been an epoch in the subject. All that is over with me; & I will only think on the admirable skill with which you have selected the striking points & explained them. . . .

I know you will forgive me for writing with perfect freedom; for you must know how deeply I respect you, as my old honoured guide & master.— I heartily hope & expect that your Book will have gigantic circulation & may do in many ways as much good as it ought to do.—

I am tired; so no more. I have written so briefly that you will have to guess my meaning. I fear my remarks are hardly worth sending— Farewell—with kindest remembrances to Lady Lyell | Ever yours | C. Darwin

[1] Charles Lyell, *The geological evidences of the antiquity of man with remarks on theories of the origin of species by variation* (London, 1863).

From Charles Lyell 11 March 1863

53 Harley Street:
March 11, 1863.

My dear Darwin,—

I see the 'Saturday Review' calls my book 'Lyell's Trilogy on the Antiquity of Man, Ice, and Darwin.'

As to my having the authority you suppose to lead a public who up to this time have regarded me as the advocate of the other side (as in the 'Principles') you much overrate my influence. In the new 'Year Book of Facts' for 1863, of Timbs, you will see my portrait, and a sketch of my career, and how I am the champion of anti-transmutation. I find myself after reasoning through a whole chapter in favour of man's coming from the animals, relapsing to my old views whenever I read again a few pages of the 'Principles,' or yearn for fossil types of intermediate grade. Truly I ought to be charitable to Sedgwick and others. Hundreds who have bought my book in the hope that I should demolish heresy, will be awfully confounded and disappointed. As it is, they will at best say with Crawfurd, who still stands out, 'You have put the case with such moderation that one cannot complain.' But when he read Huxley, he was up in arms again.

My feelings, however, more than any thought about policy or expediency, prevent me from dogmatising as to the descent of man from the brutes, which, though I am prepared to accept it, takes away much of the charm from my speculations on the past relating to such matters.

I cannot admit that my leap at p. 505, which makes you 'groan,' is more than a legitimate deduction from 'the thing that is' applied to 'the thing that has been,' as Asa Gray would say, and I have only put it moderately, and as a speculation.

I cannot go Huxley's length in thinking that natural selection and variation account for so much, and not so far as you, if I take some passages of your book separately.

I think the old 'creation' is almost as much required as ever, but of course it takes a new form if Lamarck's views improved by yours are adopted.

What I am anxious to effect is to avoid positive inconsistencies in different parts of my book, owing probably to the old trains of thought, the old ruts, interfering with the new course.

But you ought to be satisfied, as I shall bring hundreds towards you, who if I treated the matter more dogmatically would have rebelled.

I have spoken out to the utmost extent of my tether, so far as my reason goes, and farther than my imagination and sentiment can follow, which I suppose has caused occasional incongruities.

Woodward is the best arguer I have met with against natural selection and variation. He puts conchological difficulties against it very forcibly. He is at the same time an out-and-out progressionist.

I am glad that both you and Hooker like the 'ice' part of the Trilogy. You are the first to allude to my remarks on Ramsay, who says 'I shall come round to his views in good time.'

Falconer, whom I referred to oftener than to any other author, says I have not done justice to the part he took in resuscitating the cave question, and says he shall come out with a separate paper to prove this. I offered to alter anything in the new edition, but this he declined. Pray write any criticism that occurs to you; you cannot put them too strongly or plainly.

Ever yours sincerely, | Charles Lyell.

From Asa Gray 22–30 March 1863

Cambridge [Massachusetts]
22. March '63

My Dear Darwin

Argyle's article on the *Supernatural*—to which you called my attention a long while ago, I never happened to see till to-day—when I have read it through.[1] It is quite clever—not deep, but clear, and I think useful I see no occasion for finding fault with him—except in his attempts now and then to direct a little odium against you—which is unhandsome—for his main points are those I hammered out in Atlantic.[2] &c—indeed I see signs of his having read the same. But it is hardly fair of him, after expressing his complete conviction that where the operation of natural causes can be clearly traced, the implication of design—upon its appropriate evidence—is not thereby rendered less certain or less convincing.—to go on to speak of derivation-doctrines in a way that implies the contrary.

Of course we believers in *real design*, make the most of your frank and natural terms, 'contrivance, purpose", &c"—and pooh-pooh your endeavors to resolve such contrivances into necessary results of certain physical processes, and make fun of the race between long noses and long nectaries!

23^{d} March.

D^{r}. Wyman—who is a sharp fellow tells me that—on the authority of the historian Prescott,—the Incas of Peru for—no one knows how long—married *their sisters*—to keep the perfect purity of the blood. Quere. How did this strong case of close-breeding operate? Did they run out thereby? Wyman thinks there is no evidence of it.

If it is true—and the Incas stood it for a long course of generations, you must look to it—for it will bear hard against your theory of the necessity of crossing.

If they run out, you will have a good case.

P.S. *30 March.* ...

You think Lyell *too non-comittal* and *timid*. Well Huxley makes up for it, I should think!

Ever dear Darwin | Yours cordially | Asa Gray

1 [G. D. Campbell, eighth duke of Argyll], 'The supernatural', *Edinburgh Review* 116 (1862): 378–97.

2 [Asa Gray], 'Darwin on the origin of species', *Atlantic Monthly* 6 (1860): 109–16, 229–39; 'Darwin and his reviewers', *ibid.*, pp. 406–25.

[An anonymous review of William Benjamin Carpenter's *Introduction to the study of the Foraminifera* appeared in the *Athenæum* of 28 March 1863. The review derided CD for having used what it described as 'Pentateuchal terms' in *Origin* when referring to a primordial form 'into which life was first breathed' (*Origin*, p. 484). The author of the review, later identified as Richard Owen, followed the German naturalist and leading exponent of romantic *Naturphilosophie*, Lorenz Oken, in arguing that microscopic organisms like the Foraminifera were spontaneously generated on the beds of seas, lakes, and rivers, by the effect of a 'general polarizing force' on the 'slime' from dead and decaying organisms. CD rarely responded to criticism in print, but in this case he published a lengthy rejection of the review, also in the *Athenæum*.]

To *Athenæum* 18 April [1863]

Down, Bromley, Kent,
April 18.

I hope that you will permit me to add a few remarks on Heterogeny, as the old doctrine of spontaneous generation is now called, to those given by Dr. Carpenter, who, however, is probably better fitted to discuss the question than any other man in England. Your reviewer believes that certain lowly organized animals have been generated spontaneously—that is, without pre-existing parents—during each geological period in slimy ooze. A mass of mud with matter decaying and undergoing complex chemical changes is a fine hiding-place for obscurity of ideas. But let us face the problem boldly. He who believes that organic beings have been produced during each geological period from dead matter must believe that the first being thus arose. There must have been a time when inorganic elements alone existed on our planet: let any assumptions be made, such as that the reeking atmosphere was charged with carbonic acid, nitrogenized compounds, phosphorus, &c. Now is there a fact, or a shadow of a fact, supporting the belief that these elements, without the presence of any organic compounds, and acted on only by known forces, could produce a living creature? At present it is to us a result absolutely inconceivable. Your reviewer sneers with justice at my use of the "Pentateuchal terms," "of one primordial form into which life was first breathed": in a purely scientific work I ought perhaps not to have used such terms; but they well serve to confess that our ignorance is as profound on the origin of life as on the origin of force or matter.[1] Your reviewer thinks that the weakness of my theory is demonstrated because existing Foraminifera are identical with those which lived at a very remote epoch. Most naturalists look at this fact as the simple result of descent by ordinary reproduction; in no way different, as Dr. Carpenter remarks, except in the line of descent being longer, from that of the many shells common to the middle Tertiary and existing periods.

The view given by me on the origin or derivation of species, whatever its weaknesses may be, connects (as has been candidly admitted by some of its opponents, such as Pictet, Bronn, &c.) by an intelligible thread of reasoning a multitude of facts: such as the formation of domestic races by man's selection,—the classification and affinities of all organic beings,—the innumerable gradations in structure and instincts,—the similarity of pattern in the hand, wing or paddle of animals of the same great class,—the existence of organs become rudimentary by disuse,—the similarity of an embryonic reptile, bird and mammal, with the retention of traces of an apparatus fitted for aquatic respiration; the retention in the young calf of incisor teeth in the upper jaw, &c.,—the distribution of animals and plants, and their mutual affinities within the same region,—their general geological succession, and the close relationship of the fossils in closely consecutive formations and within the same country; extinct marsupials having preceded living marsupials in Australia, and armadillo-like animals having preceded and generated armadilloes in South America,—and many other phenomena, such as the gradual extinction of old forms and their gradual replacement by new forms better fitted for their new conditions in the struggle for life. When the advocate of Heterogeny can thus connect large classes of facts, and not until then, he will have respectful and patient listeners.

Dr. Carpenter seems to think that the fact of Foraminifera not having advanced in organization from an extremely remote epoch to the present day is a strong objection to the views maintained by me. But this objection is grounded on the belief—the prevalence of which seems due to the well-known doctrine of Lamarck— that there is some necessary law of advancement, against which view I have often protested. Animals may even become degraded, if their simplified structure remains well fitted for their habits of life, as we see in certain parasitic crustaceans. I have attempted to show ('Origin,' 3rd edit. p. 135) that lowly-organized animals are best fitted for humble places in the economy of nature; that an infusorial animalcule or an intestinal worm, for instance, would not be benefited by acquiring a highly complex structure. Therefore, it does not seem to me an objection of any force that certain groups of animals, such as the Foraminifera, have not advanced in organization. Why certain whole classes, or certain numbers of a class, have advanced and others have not, we cannot even conjecture. But as we do not know under what forms or how life originated in this world, it would be rash to assert that even such lowly endowed animals as the Foraminifera, with their beautiful shells as figured by Dr. Carpenter, have not in any degree advanced in organization. So little do we know of the conditions of life all around us, that we cannot say why one native weed or insect swarms in numbers, and another closely allied weed or insect is rare. Is it then possible that we should understand why one group of beings has risen in the scale of life during the long lapse of time, and another group has remained stationary? Sir C. Lyell, who has given so excellent a discussion on species in his great work on the 'Antiquity of Man,' has advanced a somewhat analogous objection, namely, that the mammals, such as seals or bats, which alone have been enabled to reach oceanic islands, have not been developed into various terrestrial forms, fitted to fill the unoccupied places in their new island-homes; but Sir Charles has partly answered his own objection. Certainly I never anticipated that I should have had to encounter objections on the score that organic beings have not undergone a greater amount of change than that stamped in plain letters on almost every line of their structure. I cannot here resist expressing my satisfaction that Sir Charles Lyell, to whom I have for so many years looked up as my master in geology, has said (2nd edit. p. 469):— "Yet we ought by no means to undervalue the importance of the step which will have been made, should it hereafter become the generally received opinion of men of science (as I fully expect it will) that the past changes of the organic world have been brought about by the subordinate agency of such causes as Variation and Natural Selection." The whole subject of the gradual modification of species is only now opening out. There surely is a grand future for Natural History. Even the vital force may hereafter come within the grasp of modern science, its correlations with other forces have already been ably indicated by Dr. Carpenter in the *Philosophical Transactions*;[2] but the nature of life will not be seized on by assuming that Foraminifera are periodically generated from slime or ooze.

Charles Darwin.

[1] In a letter to J. D. Hooker of [29 March 1863] (*Correspondence* vol. 11, CD wrote: 'Who would have ever thought of the old stupid Athenæum taking to Oken-like transcendental philosophy written in Owenian style! It will be some time before we see "slime, snot or protoplasm" (what an elegant writer) generating a new animal. But I have long regretted that I truckled to public opinion & used Pentateuchal term of creation, by which I really meant "appeared" by some wholly unknown process.—'

[2] W. B. Carpenter, 'On the mutual relations of the vital and physical forces', *Philosophical Transactions of the Royal Society* (1850): 727–57.

To Asa Gray 20 April [1863]

Down Bromley Kent
April 20th

My dear Gray.—

Thanks for two notes, one about the Duke of A[rgyll] (did I tell you how capitally this was criticised in the Saturday Review?) & the other note very savage against England: I cannot help fearing that we shall drift into war: what a curse it will be to us anyhow; for you seem to be getting to like war.— ... We must keep to science, I fear, for we both seem to be getting to think each other's country conduct worse & worse. . . .

Your remark from Wyman about the Incas came very appropriately for I am at present summing up all facts on this subject & give facts on both sides.— I now regret that I have always intentionally evaded case of man; but have put in a note on such facts as I have heard of.—

... I forgot to ask about Intermarriage; M. Devay says explicitly that state of Ohio, from evil shown by statistical returns, has legislated against cousins marrying:[1] can this be true?

Here is another heterogenous question: have you ever formed any theory, why in spire of leaves (I have been reading your most clear account in your Lessons) the angles go $\frac{1}{2}$ $\frac{1}{3}$ $\frac{2}{5}$ $\frac{3}{8}$ &c— Why should there not be $\frac{1}{4}$ or $\frac{1}{5}$? It seems to me most marvellous— There must be some explanation. If you have theory, I know it would be too long to explain; but I shd like to hear whether you have. My good friend Falconer has been twitting me that these angles go by as fixed a law as that of Gravity & *never vary*. I can fancy that packing of organs in very early bud may cause general alternation in the parts of the flower & consequent interruption in the spires.—

Speaking of Falconer, I was very sorry to see his letter in the Athenæum; so irreverent & virulent towards Lyell. We have had lately sharp sparring in the Athenæum. Did you see the article on Heterogeny or Spontaneous generation, written I believe, certainly by Owen!! it was in Review on Carpenter, who seems to have been sillily vexed at Owen calling me Carpenter's master; it was like his clever malignity. Under the cloak of a fling at Heterogeny I have sent a letter to Athenæum in defence of myself, & I take sly advantage to quote Lyells *amended* verdict on the Origin.—

I suppose my letter will appear next week: it is no great thing.— . . .

. . . I have now no doubt that you are perfectly right about fertilisation of Cypripedium; a friend lent me a plant of C. pubescens. I put excessively minute Bee (an Andrœna) into the Labellum & covered orifice with wet paper; but this precaution I afterwards found superfluous, for the edges all round of the orifice of labellum are folded over (just like insect traps for London kitchens) so that the Bee could not crawl out.— Well the bee immediately crawled out by one of the windows opposite the anthers, with his back towards the anther, against which he firmly pressed it, owing to the elastic wool opposite the anther. It was pretty to see under lens how the whole thorax & base of wings was smeared with pollen. I put him back into the labellum five times & five times I saw his back smeared. As you know he must pass under the stigma (with its spines directed towards the apex as you describe), for there is no other passage; & as I expected when I cut open the flower I found the stigma well smeared with pollen.— It was beautiful.—

Good Night | My dear Gray | Ever yours cordially | Though an Englishman | Charles Darwin

[1] Francis Devay, *Du danger des mariages consanguins sous le rapport sanitaire*, 2d edition (Paris, 1862).

To Asa Gray 11 May [1863]

Down Bromley Kent [Leith Hill Place]
May 11th

My dear Gray

I have to thank you for 2 or 3 little notes. The last I was glad to receive on Lyell, & will tell him, when I write, what you say on Species-portion. I am pleased at it; but cannot quite agree. You speak of Lyell as a Judge; now what I complain of is that he declines to be Judge. It put me into despair, when I see such men as Lyell & you incapable (as you think) of deciding: I have sometimes almost wished that Lyell had pronounced against me. When I say "me"; I mean only *change of species by descent.* That seems to me the turning point. Personally, of course, I care much about Natural Selection; but that seems to me utterly unimportant compared to question of *Creation* **or** *Modification.*— Like a huge Ass I have written two stupid letters to Athenæum: the latter to above effect.—

How clever & original & candid your remark about Language & Design.—

Your little discussion on Angles of Divergence of leaves in a Spire has almost driven me mad. My 2^{d} Boy George is a good mathematician, & when I showed him the fractions, he said they formed a converging series; & I see when protracted, they do all crowd round one point. . . . I have been drawing all the real angles & unreal angles on a spire, & I see the angles which do not occur in nature, are just as symmetrical in position as the real angles. If you wish to save me from a miserable death, do tell me why the angles of $\frac{1}{2}$ $\frac{1}{3}$ $\frac{2}{5}$ $\frac{3}{8}$ &c series occur, & no other angles.— It is enough to drive the quietest man mad.—

Did you & some mathematician publish some paper on subject; Hooker says you did. Where is it? I have been visiting for a fortnight houses of relation to try to get a little health for my youngest Boy (the Natural Selection Hero)[1] & self; with very poor success. This has led me to muddle my brains over the angles of leaves.— Do you know of any plant in which angle is fluctuating or variable? I often bless science; for when observing I forget my discomfort & at no other time am I comfortable for two successive hours.— . . .

Farewell— Yours most truly | C. Darwin

[1] In a letter to Gray of [3–]4 September 1862 (*Correspondence* vol. 10), CD had written:

> Progress of Education.— one of my little Boys Horace said to me, "there are a terrible number of adders here; but if everyone killed as many as they could, they would sting less".— I answered "of course they would be fewer" Horace "Of course, but I did not mean that; what I meant was, that the more timid adders, which run away & do not sting would be saved, & after a time none of the adders would sting".—
>
> Natural selection!!

To J. D. Hooker   15 and 22 May [1863]

Down Bromley Kent.
May 15th

My dear Hooker

Your letter received this morning interested me more than even most of your letters; & that is saying a good deal.—

I must scribble a little on several points. About Lyell & Species you put the *whole* case, I do believe, when you say that he is "half-hearted & whole-headed". I wrote to A. Gray that when I saw such man as Lyell & he refuse to judge, it put me in despair; & that I sometimes thought I sh^d. prefer that Lyell had judged against modification of species, rather than profess inability to decide; & I left him to apply this to himself.— I am heartily rejoiced to hear that you intend to try to bring L. & F. together again: but had you not better wait till they are a little cooled? you will do science a real good service. Falconer never forgave Lyell for taking the Purbeck bones from him & handing them over to Owen.—[1] ...

With respect to Island Floras, if I understand rightly, we differ almost solely how plants first got there: I suppose that at long intervals, from as far back as later Tertiary periods, to the present time plants occasionally arrived (in some cases perhaps aided by different current from existing currents & by former islands) & that the old arrivals have survived little modified on the islands, but have been greatly modified or become extinct on the continents. If I understand, you believe that all islands were formerly united to continents & then received all their plants & none since; & that on the islands they have undergone less extinction & modification than on the continents.— The number of animal-forms on islands very closely allied to those on continents, with a few extremely distinct & anomalous, does not seem to me well to harmonise with your supposed view of *all* having formerly arrived or rather having been left together on the island.— ...

All my work is at wretched standstill, with everlasting sickness & devilish headachs—

Goodnight | My dear old friend. | C. Darwin ...

[1] Bones of early fossil mammals from the middle Purbeck beds at Durlstone Bay near Swanage, Dorset, were excavated in the 1850s, the first find being described by Richard Owen. In 1857, Charles Lyell was instrumental in encouraging Samuel Husband Beckles to undertake further excavations. Beckles sent the fossils directly to Lyell, and Lyell transferred them to Hugh Falconer, who made the initial determinations. It was intended that when the collection was completed the fossils would be transferred to Owen for description and publication. Lyell apparently did this earlier than had been arranged. Owen published descriptions of the fossils in 1871.

To George Bentham 22 May [1863]

Down. | Bromley. | Kent. S.E.
May 22[d]

My dear Bentham

I am much obliged for your kind & interesting letter. I have no fear of anything that a man like you, will say annoying me in the very least degree. On the other hand any approval from one, whose judgment & knowledge I have for many years so sincerely respected, will gratify me much.— The objection, which you well put, of certain forms remaining unaltered through long time & space, is no doubt formidable in appearance & to a certain extent in [reality] according to my judgment.— But does not the difficulty rest much on our silently assuming that we know more than we do? I have literally found nothing so difficult as to try & always remember our ignorance. I am never weary when walking in any new adjoining district or country of reflecting how absolutely ignorant we are why certain old plants are not there present, & other new ones are & others in different proportions. If we once fully feel this, then in judging the theory of natural selection, which implies that a form will remain unaltered unless some alteration be to its benefit, is it so very wonderful that some forms should change much slower & much less, & some few should have changed not at all under conditions which to us (who really know nothing what are the important conditions) seem very different.— Certainly a priori we might have anticipated that all the plants anciently introduced into Australia would have undergone some modification; but the fact that they have not been modified does not seem to me a difficulty of weight enough to shake a belief grounded on other arguments.— I have expressed myself miserably, but I am far from well today.—

I am very glad that you are going to allude to Pasteur; I was struck with infinite admiration at his work.—

With cordial thanks, believe me Dear Bentham | Yours very sincerely | Ch. Darwin

In fact the belief in natural selection must at present be grounded entirely on general considerations. (1) on its being a vera causa, from the struggle for existence; & the certain geological fact that species do somehow change (2) from the analogy of change under domestication by man's selection. (3) & chiefly from this view connecting under an intelligible point of view a host of facts.—

When we descend to details, we can prove that no one species has changed: nor can we prove that the supposed changes are beneficial which is the groundwork of the theory. Nor can we explain why some species have changed & others have not. The latter case seems to me hardly more difficult to understand precisely & in detail than the former case of supposed change. Bronn may ask in vain the old creationist school & the new school why one mouse has longer ears than another mouse—& one plant more pointed leaves than another plant.—

[The following letter has been translated from the original French. Jacques Boucher de Perthes had reported on a human jaw found in March 1863 at the Moulin-Quignon quarry, near Abbeville, France, together with teeth and some artefacts. The jaw was believed by Boucher de Perthes, Armand de Quatrefages, and other French naturalists to date from the Post-Pliocene (Pleistocene) period. However, on examining the finds in April 1863, Joseph Prestwich, John Evans, and Hugh Falconer concluded that they were forgeries. Between 9 and 13 May, a committee of French and English naturalists met in Paris to discuss the question, and concluded that the implements and jaw were authentic, although several scientists remained unconvinced.]

From Jacques Boucher de Perthes 23 June 1863

Abbeville
23 June 1863

My dear and honored Sir

Your letter of the 16th has given me great satisfaction and I thank you for it. I will always hold it a great honour to be in correspondence with you who have obtained such a fine place in Science and whose ideas approach so closely to my own.

I do not know if you were aware of my book, which appeared in 1838 under the title *de la creation, essai sur la progression des etres*. It has been mostly ignored, because it did not read as an amusing book, but such as it is—I ask you to accept it. . . .

That unfortunate jaw has caused me much trouble following the discussions to which it has given rise. It was the same 28 years ago when I exhibited my first axes. Nevertheless then as today I was certain of whatever I proposed on these matters[;] I guessed at nothing and when I assert something it is because I have seen it. I said a very long time ago that fossil man would be found and I repeat my prediction. It will be the same as those axes that I found everywhere in England as in France, in short everywhere one looks for them. Fossil man is also there under our feet, but he is not wanted. As soon as a specimen is discovered, people seek to quarrel with it. The Moulin Quignon find has given rise to so many incredible stories!— people wanted it to have been buried by hoaxers although it was acknowledged that this was physically impossible because there was too much earth at its exhumation to render a fraud practicable, and moreover, unless one is blind or senseless, one can always see when a site has been disturbed Even the least experienced workman is not fooled, so then the axes became a target. It was said that they were forgeries— well! one is no truer than the other— It was possible to manufacture some axes at St Acheul which are fairly similar to those found in peat bogs, which are indeed fairly easily imitated, but this is not the case at Moulin Quignon. Their imitation would require a great deal of skill, care and time, so that in order to make a profit the workmen would have to sell them for a great deal of money, and the normal price is 25 centimes (five sous). I wanted to be absolutely sure of this. I set up a competition in which I offered 20 Francs to the workman who could make one similar enough to those of Moulin Quignon to fool people. After very many tries, no one has come near to making even a passable axe.

Please accept, my dear sir, the expression of my high esteem and complete devotion. | J. Boucher de Perthes …

[In August 1863, English newspapers were predicting that Washington, D.C., was about to fall following the invasion of Maryland and Pennsylvania by the Confederate army. At the same time, there had been a revival of interest in the history of the Crimean War (1853–6) following the publication of the first volume of Alexander William Kinglake's *The invasion of the Crimea* (Edinburgh and London, 1863–87).]

To Asa Gray 4 August [1863]

Down Bromley Kent
Aug 4th

My dear Gray. . . .

How profoundly interesting American new[s] is— I declare almost more so even to us than Crimean new[s].— Do not hate poor old England too much. Anyhow she is the mother of fine children all over the world. I declare no man could have tried to wish more sincerely for the north tha[n] I have done.— My reason tells me that perhaps it would be best,—of course best if it would end Slavery, but I cannot pump up enthusiasm. The boasting of your newspapers & of your little men, & the abuse of England, and the treatment of the free coloured population, and the not freeing Maryland slaves[1] stops all my enthusiasm. If all the States were like New England the case would be different.— I find a man cannot hope by intention. You will think me a wretched outcast. Farewell & do not hate me much. What devils the low Irish have proved themselves in New York. If you conquer the South you will have an Ireland fastened to your tail.—[2]

Good night & Farewell | C. Darwin

[1] President Abraham Lincoln's Emancipation Proclamation, which came into effect on 1 January 1863, had freed the slaves only in states rebelling against the Union. Since Maryland had not seceded from the Union, slaves were not freed in that state.

[2] In July 1863, Irish immigrants had played a prominent role in the riots in New York City against the drafting of men into the Union army; the riots resulted in the deaths of at least 105 people.

From Emma Darwin to Patrick Matthew 21 November [1863]

Down. | Bromley. | Kent. S.E.
Nov 21.

Dear Sir

Mr Darwin begs me to thank you warmly for your letter which has interested him very much. I am sorry to say that he is so unwell as not to be able to write himself.

With regard to Natural Selection he says that he is not staggered by your striking remarks. He is more faithful to your own original child than you are yourself.[1] He says you will understand what he means by the following metaphor.

Fragments of rock fallen from a lofty precipice assume an infinitude of shapes—these shapes being due to the nature of the rock, the law of gravity &c— by merely selecting the well-shaped stones & rejecting the ill-shaped an architect (called Nat. Selection[)] could make many & various noble buildings.

Mr Darwin is much obliged to you for sending him your photograph. He wishes he could send you as good a one of himself. The enclosed was a good likeness taken by his eldest son but the impression is faint.

You express yourself kindly interested about his family. We have 5 sons & 2 daughters, of these 2 only are grown up. Mr Darwin was very ill 2 months ago & his recovery is very slow, so that I am afraid it will be long before he can attend to any scientific subject.

Dear Sir | yours truly | E. Darwin

[1] See letter to *Gardeners' Chronicle*, [13 April 1860].

1864

To A. R. Wallace 1 January 1864

Down. | Bromley. | Kent. S.E.
Jan 1. 1864

Dear Wallace

I am still unable to write otherwise than by dictation. In a letter received 2 or 3 weeks ago from Asa Gray he writes "I read lately with *gusto* Wallace's exposé of the Dublin man on Bee cells &c"[1]

Now tho' I cannot read at present I much want to know where this is published that I may procure a copy. Further on Asa Gray says (after speaking of Agassiz's paper on Glaciers in the Atlantic Magazine) & his recent book entitled Method of Study[2] "Pray set Wallace upon these articles" So Asa Gray seems to think much of your powers of reviewing & I mention this as it assuredly is laudari a laudato.[3] I hope you are hard at work & if you are inclined to tell me I sh[d] much like to know what you are doing. It will be many months I fear before I shall do any thing.

Pray believe me yours very sincerely. | Ch. Darwin

1 Samuel Haughton had attacked CD's argument that the hive-bee's cells provided evidence for natural selection. Alfred Russel Wallace's criticism appeared in his 'Remarks on the Rev. S. Haughton's paper on the bee's cell, and on the origin of species', *Annals and Magazine of Natural History* 3d ser. 12 (1863): 303–9.

2 Louis Agassiz, 'The formation of glaciers', *Atlantic Monthly* 12 (1863): 568–76, and *Methods of study in natural history* (Boston, Mass., 1863).

3 To be praised by one who is himself praised.

From A. R. Wallace 2 January 1864

5, Westbourne Grove Terrace, W.
Jan 2nd. 1864

My dear Darwin

Many thanks for your kind letter. I was afraid to write because I heard such sad accounts of your health but I am glad to find that you can write & I presume read, by deputy. My little article on Haughton's paper was published in the "Annals of Nat. Hist." about Aug. or Sept. last I think, but I have not a copy to refer to. I am sure it does not deserve Asa Grays praises for though the matter may be true enough, the manner I know is very inferior. It was written hastily & when I read it in the "Annals" I was rather ashamed of it as I knew so many could have done it so much better.

I will try & see Agassiz' paper & book. What I have hitherto seen of his on glacial subjects seemed very good, but in all his Nat. Hist. *theories*, he seems so utterly wrong & so totally blind to the plainest deductions from facts, & at the same time so vague & obscure in his language, that it would be a very long & wearisome task to answer him.

With regard to work I am doing but little— I am afraid I have no good habit of systematic work. I have been gradually getting parts of my collections in order, but the obscurities of synonomy & descriptions, the difficulty of examining specimens & my very limited library, make it wearisome work. I have been lately getting the first groups of my butterflies in order, & they offer some most interesting facts in variation & distribution,—in variation some very puzzling ones— Though I have very fine series of specimens I find in many cases I want more, in fact if I could have afforded to have had all my collections kept till my return I should I think have found it necessary to retain twice as many as I now have.

I am at last making a beginning of a small book on my Eastern journey, which if I can persevere I hope to have ready by next 'Xmas. I am a very bad hand at writing anything like narrative, I want something to argue on & then I find it much easier to go a'head. I rather despair therefore of making so good a book as Bates', though I think my subject is better. Like every other traveller I suppose, I feel dreadfully the want of copious notes on common every day objects, sights & sounds & incidents, which I imagined I could never forget but which I now find it impossible to recall with any accuracy. . . .[1]

Allow me to say in conclusion how much I regret that unavoidable circumstances have caused me to see so little of you since my return home, & how earnestly I pray for the speedy restoration of your health.

Yours most sincerely | Alfred R. Wallace

[1] H. W. Bates, *The naturalist on the River Amazons. A record of adventures, habits of animals, sketches of Brazilian and Indian life, and aspects of nature under the equator, during eleven years of travel* (London, 1863). Wallace did eventually publish his *The Malay Archipelago: the land of the orang-utan, and the bird of paradise* (London, 1869).

To J. D. Hooker [27 January 1864]

Down—
Wednesday

My dear old Friend.

I was very glad to get your last letter which crossed mine on the road. It told me a lot of news; for I hear from no one else. How good you are.— Nothing would please me more than to see you here, if you had time; but as yet it would be very rash in me, as it surely w^{d}. bring on my vomiting, & I sh^{d}. suppose few human beings had vomited so often during the last 5 months. For several days I have been decidedly better, & what I lay much stress on (whatever Doctors say) my brain feels far stronger & I have lost many dreadful sensations.— The Hot-House is such an amusement to me; & my amusement I owe to you, as my delight is to look at the many odd leaves & plants from Kew.— Ceropegia Gardeneri is now in flower, & I think it the oddest flower I ever saw. Do you know it? with the points of the corolla stictched together in centre,—to keep out big insects I say.

The only approach to work which I can do is to look at tendrils & climbers, this does not distress my weakened Brain— …

Farewell my dear old friend | C. Darwin …

To Asa Gray 25 February [1864]

Down Bromley Kent
Feb. 25

My dear Gray

You have been so kind & good a friend to me, that I think you will like to have a note in pencil to hear that I am better. The vomiting is not now daily & on my good days, I am much stronger. My head hardly now troubles me, except singing in ears— It is now six months since I have done a stroke of work; but I begin to hope that in a few more months, I may be able to work again.— I am able most days now to get to my Hot-house I amuse myself a little by looking at climbing plants. The first job which I shall do is to draw up result of Lythrum crosses & on movements of climbing plants.—

I have of course seen no one & except good dear Hooker, I hear from no one. He like a good & true friend, though so overworked, often writes to me.—

I have had one letter which has interested me greatly with a paper which will appear in Linn. Journal by Dr. Cruger of Trinidad, which shows that I am all right about Catasetum.[1] Even to spot where pollinia adhere to Bees, which visit flower, as I said, to gnaw the labellum.— Cruger's account of Coryanthes & the use of the bucket-like labellum full of water beats everything: I *suspect* the Bees being well wetted flattens hairs & allows viscid disc to adhere.

I have given up hearing the newspaper read aloud as Books are more amusing & less tiring. Good Heavens the lot of trashy novels, which I have heard is astounding.— I have heard little about America.— You wrote me some little time ago a pleasant letter, which for a month I have been wishing to answer & thank you for.— Sometime let me hear what you are doing & what you expect for your country.

Your poor broken down brother naturalist & affetiont friend, | C. Darwin

I wish Dana was not so imaginative & speculative in his writings.

[1] Hermann Crüger confirmed CD's observations in 'Three sexual forms of *Catasetum tridentatum*'.

[Early in March 1864, John Scott sent CD a short note announcing that he had left his position at the Royal Botanic Garden, Edinburgh. CD, had asked Scott for an explanation. This letter is Scott's response.]

From John Scott 28 March 1864

Denholm
28th. Mar. 1864.

Sir. . . .

I am sorry to trouble you—in your present weak state of health—with anything concerning my future plans; but as you have so kindly desired me, I will do so as briefly as possible.

In respect to future plans, I regret to state that I have no definite plans whatever to look to.— I left the Bot. Gard. Edinburgh completely chagrined with my masters behaviour to me.— Though there have been several young men sent out to India by Prof. Balfour & M^r. M^c.Nab since I refused to accept the situation they offered me at Darjeeling—respecting which perhaps you may remember I consulted you—they have never given me another offer. (And I know there has been one or two really good places at their disposal). I felt this repeated overlooking very deeply, as I had been led to hope that they would do what they could to get me some foreign situation.

Seeing, however, the utter hopelessness of entertaining this further, I with the completion of the experiment, in which I was immediately engaged intimated to M^r. M^c.Nab my intention to leave, which I did at the conclusion of my 5th. year in his service.

I would be glad indeed, if I might yet be permitted to entertain the hope that D^r. Hooker will 'remember me when any place likely to suit, comes under his notice'—as he was once induced to write by the kind interest you were pleased to honour me with.

Sincerely regretting that I should have had any occasion for troubling you—more especially at present—with these merely personal details | I remain | Sir | J. Scott

[Alfred Newton had made available to CD the foot of a red-legged partridge with a ball of earth containing seeds attached; he had also exhibited the foot at a meeting of the Zoological Society of London on 21 April 1863 'as a singular illustration of the manner in which birds may occasionally aid in the dispersion of seeds'.]

To Alfred Newton 29 March [1864]

Down, Bromley | Kent
March 29th.

My dear Sir

Since receiving your letter of Oct. 21st., I have been, & am still ill; but I managed to examine the partridges leg— the toes, & tarsus were frightfully diseased, enlarged & indurated. There were no concentric layers in the ball of earth, but I cannot doubt that it had become slowly aggregated, probably the result of some viscid exudation from the wounded foot. It is remarkable, considering that the ball is 3 years old, that 82 plants have come up from it—12 being monochot. & 70 dichot. consisting of at least 5 different plants—perhaps many more. The bird limping about during the autumn would easily collect many seeds on the viscid surface. I am extremely much obliged to you for sending me this interesting specimen.

I am, dear Sir | Yours very faithfully | Charles Darwin

To Louis Agassiz 12 April 1864

Down. | Bromley. | Kent. S.E.
Ap. 12. 1864

My dear Sir

Owing to long continued illness & absence from London, I received only a few days ago the copy of your "Methods of Study" with some other publications, & your kind note of introduction to Mr Lesly.

I thank you sincerely for the above present.

I know well how strongly you are opposed to nearly everything I have written & it gratifies me deeply that you have not for this cause taken, like a few of my former English friends, a personal dislike to me.

With my cordial thanks & sincere respect | I remain my dear Sir | yours very sincerely | Charles Darwin

From Benjamin Dann Walsh 29 April – 19 May 1864

Rock. Island. Illinois. U.S.
April 29. 1864

Chas Darwin Esq.
Dear Sir,

More than thirty years ago I was introduced to you at your rooms in Christ's College by A. W. Griesbach & had the pleasure of seeing your noble collection of British Coleoptera. Some years afterwards I became Fellow of Trinity & finally gave up my fellowship, rather tha[n] go into Orders, & came to this Country. For the last 5 or 6 years I have been paying considerable attention to the insect Fauna of U.S., some of the fruits of which you will see in the enclosed pamphlets.

Allow me to take this opportunity of thanking you for the publication of your Origin of Species, which I read three years ago by the advice of a Botanical friend, though I had a strong prejudice against what I supposed then to be your views. The first perusal staggered me, the second convinced me, & the oftener I read it the more convinced I am of the general soundness of your theory.

As you have called upon Naturalists that believe in your views to give public testimony of their convictions, I have directed your attention on the outside of one or two of my Pamphlets to the particular passages in which ⟨I⟩ have done so. You will please accept these Papers from me in token of my respect & admiration.

As you may see from the latest of these Papers, I ⟨have⟩ recently made the remarkable discovery that there ⟨are the⟩ so-called "three sexes" not only in social insects but ⟨also in the⟩ strictly solitary genus *Cynips*. . . .

Very truly yours, Benj. D. W⟨alsh⟩ . . .

[John Scott had been without employment since leaving his position as foreman of the propagating department at the Royal Botanic Garden, Edinburgh, in March 1864. Scott had corresponded regularly with CD since 1862, and CD suspected that his encouragement of Scott's botanical researches might have contributed to his difficulties with his former employers.]

From J. D. Hooker   19 May 1864

Kew
May 19th/64.

Dear Darwin

I have been thinking a great deal about Scott, & have quite come to the conclusion that India would be the place for him. Anderson of Calcutta writes me word that the Govt. are about to settle the Forest management for Bengal; & this will afford plenty of opportunity for good men to get on. Indeed what with Tea Cinchona & Indigo & Coffee, a man of proved probity can have no difficulty in getting on; such positions offer abundant means of following any pursuit, & Scotts temper would be no objection.

He would of course want assistance to get out to India for it is a mere chance that such places are advertized for in this country & passage out paid.— but once out there, with good introductions, & I am sure such a man should do well. I can, if you like, write out to Anderson at Calcutta, to Cleghorn Inspector of Forests of N.W. India & Beddome, ditto of Madras, & Scott might go out to Calcutta with all the further introductions he could get. I am sure that Anderson would give or find him inexpensive quarters at the Bot. Gardens for some weeks. India is now the place of all others for active & energetic men.

I saw Herbert Spencer two days ago[;] he tells me that he has been reading Zoonomia, & intends to give all the credit hitherto awarded Lamarck to your Grdfather—[1] by the way I saw a very good portrait of the old gentleman at the Lichfield Museum.—

. . .

I wish you would soon publish a note on tendrils &c, or you will certainly be cut out by some foreigner:— A few lines to the Gard[ener's] Chronicle would suffice I am delighted to hear that you are at *Lythrum*. . . .

Ever yrs affec | J D Hooker

[1] Erasmus Darwin, author of *Zoonomia; or, the laws of organic life* (London, 1794–6), had published evolutionary ideas anticipating those of Jean Baptiste de Lamarck. Herbert Spencer was working on *Principles of biology* (London, 1864–7); he discussed Erasmus Darwin and Lamarck in an instalment issued in October 1864, giving both credit for their contributions to evolutionary theory.

To John Scott 21 May [1864]

Down, Bromley, Kent,
May 21st.

Dear Sir,—

I received from my good friend Dr Hooker a letter of which the enclosed is an extract. You had better deliberately consider what he suggests and consult your friends. Remember that Dr H. knows India, and is acquainted with many men who are now in India, and who have been there. The suggestion comes entirely from him, and was not first made by me. Reflect well, for it is an important step for you, and I do not like to take the responsibility of giving advice. If you decide to try the plan and run such risk as there is of not getting employment, can you get a character for probity, sobriety and energy, from Professor Balfour, Mr Macnab, or any clergyman or magistrate of the district in which you reside. These would be of important service. I am a little doubtful whether your scientific attainments ought to be much insisted on, though they should be mentioned. The expense of some outfit for the voyage itself, and of giving you means to subsist for a short time in India, would be considerable, but how much I do not at all know: could you enquire from any gardener who has gone out to India? If your friends approve, have they the power to assist you. I would gladly pay half, and if your friends cannot assist you I am quite ready to pay the whole, for I am sure you would put me to no unnecessary expense. You will be wrong to feel any scruple in accepting this offer on my part, for I can afford it, and it will in every way give me satisfaction both as helping you and as forwarding science. You would have time, if you accept, to finish your papers during the voyage, and I would see to their publication.

With every wish for you to decide best for yourself, | I remain, yours very faithfully, Dear Sir, | Charles Darwin.

To Asa Gray 28 May [1864]

Down Bromley Kent
May 28th

My dear Gray

Your kindness will make you glad to hear that I am nearly as well as I have been of late years, though a good deal weaker. I have been slowly writing a paper on Lythrum, & this has disinclined me for the exertion of writing letters. It has been so pleasant doing a little work after 8 months inaction. Speaking of Lythrum reminds me to say that your Nesæas are looking very healthy, & Mitchella moderately so. Some time ago I received Dr. Wrights letter about Orchids: if you write to him, beg him to note what attracts insects to Begonias? do they gnaw or penetrate the petals? Also, but I care less, what attracts them to Melastomas? Poor Dr. Cruger of Trinidad, who promised to observe, is dead.—

Whenever my Lythrum paper is printed I will of course send you a copy, & I shall like to hear whether you think it as curious a case as I do.— I have got another new **sub**-class of dimorphic plants.—

An Irish nobleman on his deathbed declared that he could conscientiously say that he had never throug[h]out life denied himself any pleasure; & I can conscientiously say that I have never scrupled to trouble you.— So here goes.— Have you travelled south, & can you tell me, whether the trees, which Bignonia capreolata climbs, are covered with moss, or filamentous lichen or Tillandsia; I ask because its tendrils abhor a simple stick, do not much relish rough bark, but delight in wool or moss. They adhere in curious manner, by making little disks at end of each point, like the Ampelopsis; when the disk sticks to bundle of fibres, these fibres grow between them & then unite, so that the fibres of wool end by being embedded in middle: By the way I will enclose some specimens & if you think it worth while you can put them under the simple microscope. It is remarkable how specially adapted some tendrils are; those of Eccremocarpus scaber do not like a stick, will have nothing to say to wool; but give them a bundle of culms of grass or bundle of bristles & they seize them well.—

I have been reading with great interest von Mohls paper in Bot. Zeitung on imperfect self-fertile flowers. He quotes you that perfect flowers of Voandzeia are quite sterile— How is this known, for is it not a Madagascar plant? I presume you know that wild plants of Amphicarpea are generally sterile. How I sh[d]. like to have seed to ascertain whether this plant is sterile when fertilised.— What a curious analogous case is that of Leersia: I have just got plants of this grass.— Please remember, if you ever come across it, seeds of the Campanula perfoliata.—

Lastly (God forgive me) can you tell me whether any of your Hollies are in state of Thyme viz some plants hermaphrodites & some Females: I have a dimorphic case which, I think, will show how this state of Thyme &c arises.— . . .

. . . As for public news I am much in arrear, for I gave up for months hearing the newspaper, as I found it more fatiguing than novels. I have heard during late 9 months an astounding number of love scenes.—

What dreadful carnage you have just recently suffered.—[1] What will the end be? Will slavery perish, if so the cost is not too dear?

Farewell my dear & good friend. You will see that I have regained my ten-horse-interrogoratory-power:—farewell— yours most sincerely | C. Darwin

I send a Photograph of myself with my Beard. Do I not look venerable?

[1] Between 5 and 12 May 1864 the Union forces reported 32,000 men killed, wounded, and missing, and the Confederate forces lost an estimated 18,000 men in the battles of the Wilderness and Spotsylvania in Virginia during the American Civil War.

To A. R. Wallace 28 [May 1864]

Down Bromley Kent
March 28th[1]

Dear Wallace

I am so much better that I have just finished paper for Linn. Soc;[2] but as I am not yet at all strong I felt much disinclination to write & therefore you must forgive me for not having sooner thanked you for your paper on man received on the 11th.—[3] But first let me say that I have hardly ever in my life been more struck by any paper than that on Variation &c &c in the Reader.[4] I feel sure that such papers will do more for the spreading of our views on the modification of species than any separate Treatises on the simple subject itself. It is really admirable; but you ought not in the Man paper to speak of the theory as mine; it is just as much yours as mine. One correspondent has already noticed to me your "high-minded" conduct on this head.

But now for your Man paper, about which I shd. like to write more than I can. The great leading idea is quite new to me, viz that during late ages the mind will have been modified more than the body; yet I had got as far as to see with you that the struggle between the races of man depended entirely on intellectual & *moral* qualities.—

The latter part of paper I can designate only as grand & most eloquently done.— I have shown your paper to 2 or 3 persons who have been here & they have been equally struck with it.— I am not sure that I go with you on all minor points: when reading Sir G. Greys account of constant battles of Australian savages,[5]

I [remember] thinking that N. Selection would come in, & likewise with Esquimaux with whom the art of fishing & managing canoe is said to be hereditary. I rather differ on the rank under classificatory point of view which you assign to man: I do not think any character simply in excess ought ever to be used for the higher divisions.— Ants would not be separated from other Hymenopterous insects however high the instinct of the one & however low the instincts of the other.—

With respect to the differences of race, a conjecture has occurred to me that much may be due to the correlation of complexion (& consequently Hair) with constitution. Assume that a dusky individual best escaped miasma & you will readily see what I mean: I persuaded the Director Gen. of the Med. depart. of the army to send printed forms to the surgeons of all Regiments in Tropical countries to ascertain this point, but I dare say I shall never get any returns. Secondly I suspect that a sort of sexual selection has been the most powerful means of changing the races of man. I can shew that the diff^{t} races have a widely diff^{t} standard of beauty. Among savages the most powerful men will have the pick of the women & they will generally leave the most descendants.

I have collected a few notes on man but I do not suppose I shall ever use them. Do you intend to follow out your views, & if so would you like at some future time to have my few references & notes? I am sure I hardly know whether they are of any value & they are at present in a state of chaos. There is much more that I sh^{d} like to write but I have not strength

Believe me dear Wallace | Yours very sincerely | Ch. Darwin

Our aristocracy is handsomer (more hideous according to a Chinese or Negro) than middle classes from pick of women; but oh what a scheme is primogeniture for destroying N. Selection.—

I fear my letter will be barely intelligible to you—

[1] CD misdated the letter.

[2] 'Three forms of *Lythrum salicaria*'.

[3] A. R. Wallace, 'The origin of human races and the antiquity of man deduced from the theory of natural selection', *Anthropological Review* 2 (1864): clviii–clxx.

[4] CD refers to an abstract in the *Reader* of Wallace's paper 'On the phenomena of variation and geographical distribution as illustrated by the Papilionidæ of the Malayan region', *Transactions of the Linnean Society of London* 25 (1865–6): 1–71.

[5] George Grey, *Journals of two expeditions of discovery in north-west and western Australia, during the years 1837, 38, and 39* (London, 1841).

From A. R. Wallace 29 May [1864]

5, Westbourne Grove Terrace, W.
May 29th.

My dear Darwin

You are always so ready to appreciate what others do, & especially to over-estimate my desultory efforts, that I can not be surprised at your very kind & flattering remarks on my papers. I am glad however that you have made a few critical observations & am only sorry you were not well enough to make more, as that enables me to say a few words in explanation—

My great fault is haste. An idea strikes me, I think over it for a few days, & then write away with such illustrations as occur to me while going on— I therefore look at the subject almost solely from one point of view. Thus, in my paper on "Man" I aim solely at showing that brutes are modified in a *great variety* of ways by "Natural Selection", but that in *none of these particular ways* can man be modified, because of the superiority of his intellect.— I therefore no doubt overlook a few smaller points in which Nat. Selec. may still act on *men* & *brutes* alike. *Colour* is one of them & I have alluded to this in correlation to constitution, in an abstract I have made at Sclater's request for the Nat. Hist. Review.[1] At the same time there is so much evidence of migrations & displacements of races of man, & so many cases of peoples of distinct physical characters inhabiting the same or similar regions, & also of races of uniform physical characters inhabiting widely dissimilar regions,—that the external characteristics of the chief races of man must I think be *older* than his present geographical distribution,—& the modifications produced by correlation to favourable variations of constitution be only a secondary cause of external modification— I hope you may get the returns from the Army. They would be very interesting, but I do not expect the results would be favourable to your view.

With regard to the constant battles of savages leading to selection of physical superiority, I think it would be very imp[er]fect & subject to so many exceptions & irregularities that it could produce no *definite* result. For instance,—the strongest & bravest men would lead, & expose themselves most, & would therefore be most subject to wounds & death.— And the physical energy which led to any one tribe delighting in war, might lead to its extermination by inducing quarrels with all surrounding tribes & leading them to combine against it. Again superior cunning, stealth & swiftness of foot, or even better weapons would often lead to victory as well as mere physical strength. Moreover this kind of more or less perpetual war goes on among *all savage* peoples— It could lead therefore to no differential characters but merely to the keeping up of a certain average standard of bodily & mental health & vigour. So with selection of variations adapted to special habits of life as fishing, paddling, riding, climbing &c. &c. in different races, no doubt it must act to some extent, but will it be ever so *rigid* as to induce a definite physical modification, & can we imagine it to have had any part in producing the distinct races that now exist?

The sexual selection you allude to will also I think have been equally uncertain in its results— In the very lowest tribes there is rarely much polygamy & women are more or less a matter of purchase— There is also little difference of social condition & I think it rarely happens that any healthy & un-deformed man remains without wife & children. I very much doubt the often repeated assertion that our aristocracy are more beautiful than the middle classes. I allow that they present *specimens* of the highest kind of beauty, but I doubt the *average.*

I have noticed in country places a greater average amount of good looks among the middle classes, & besides we unavoidably combine in our idea of beauty, intellectual expression & refinement of *manners*, which often make the less appear the more beautiful. Mere physical beauty,—that is, a healthy & regular development of the body & features approaching to the *mean* or *type* of European man,—I believe is quite as frequent in one class of society as the other & much more frequent in rural districts than in cities.

With regard to the rank of man in Zoological Classification, I fear I have not made myself intelligible. I never meant to adopt Owen's or any other such views—but only to point out that from *one point* of view he was right— I hold that a distinct *family* for Man, as Huxley allows, is all that can possibly be given him *Zoologically*. But at the same time if my theory is *true*,—that while the animals which surrounded him have been undergoing modification in *all* parts of their bodies to a *generic* or even *family* degree of difference, he has been changing almost wholly in the brain & head,—then, in geological antiquity the *species* man may be as old as many mammalian *families*,—& the origin of the *family* man may date back to a period when some of the *orders* first originated—

As to the theory of "*Natural Selection*" itself, I shall always maintain it to be actually yours & your's only. You had worked it out in details I had never thought of, years before I had a ray of light on the subject, & my paper would never have convinced anybody or been noticed as more than an ingenious speculation, whereas your book has revolutionized the study of Natural History, & carried away captive the best men of the present Age. All the merit I claim is the having been the means of inducing *you* to write & publish at once. . . .

I think I made out every word of your letter though it was not always easy.

Believe me | My dear Darwin | Yours very Sincerely | Alfred R. Wallace

[1] Wallace's original paper was 'The origin of human races and the antiquity of man deduced from the theory of "natural selection"', *Anthropological Review* 2 (1864): clviii–clxx; his abstract was published in *Natural History Review* n.s. 4 (1864): 328–36.

From John Scott 29 July [1864]

Denholm
29th. July.

Sir.

I duly received your letter of the 26th. I am now almost afraid that for all the time I have had to prepare for my voyage, I might yet hear from the agent ere I was quite ready to leave. I will now state to you how this has occurred. I have been of late led to hope that a friend of mine here would advance a sum of money on my behalf, and accordingly was trusting in such, feeling ashamed to draw more deeply on your generous offers for my immediate wants. I stated to him plainly my circumstances & hopes of success in India, when I would repay him all that he might advance—asking him to assist me: but I have been disappointed, as he told me from certain alterations which he was about to make, he really could not at the present time spare me any money.

I have thus deferred mentioning to you any want of money, until I have given myself I fear little to time to again receive aid from you ere hearing from the agent as to sailing. I assure you I feel deeply in so heavily taxing your generosity—even though you have so repeatedly pressed me to let you know as to my pecuniary wants. These indeed would have been materially lessened had I not been so long out of a situation. I sincerely hope, therefore, that you will on this account excuse me in asking you for additional assistance to clear expenses here—and further believe that I am not unnecessarily drawing on your liberality.

I would therefore be greatly obliged by your transmitting to me a cheque for the amount which you think I might require on arriving in Calcutta, and also if you would include a little for sundry items & expenses here. You would thus enable me to meet all which I may incur in this country, & I fear not from the recomendations which you have induced Dr. Hooker to give me that I will soon find something to do in India.

Permit me again to assure you that I feel very sorry that I should have occasion to ask you for further pecuniary assistance here: and I will only add that whatever you are pleased to grant me now, as well as that which you have already laid out on my account will be fully & pleasently repaid so soon as I am fairly settled & succeeding in India.

With this hope, I would thus respectfully solicit your aid, and beg you to excuse my not stating any sum, And now with best thanks for your kindness to me | I have the honour to remain | Sir | Your obedt. & obliged Servt. | John Scott

P.S. I suppose a few standard works on Nat. Hist. will be all the books that I need take with me.

To Ernst Haeckel [after 10] August – 8 October [1864]

Down. | Bromley. | Kent. S.E.

Aug. Oct 8th.

Dear Sir

I thank you sincerely for your letter & the confidence you repose in me. I have been deeply interested in what you say about your poor wife. Her expression in the photograph is charming. I can to a certain extent understand what your feelings are, for I am fortunate enough to know what a treasure a wife can be & no one thought is so painful to me as the possibility of surviving her.

As you seem interested about the origin of the "Origin" & I believe do not say so out of mere compliment, I will mention a few points. When I joined the "Beagle" as Naturalist I knew extremely little about Natural History, but I worked hard. In South America three classes of facts were brought strongly before my mind: 1^{stly} the manner in which closely allied species replace species in going Southward.

2^{ndly} the close affinity of the species inhabiting the Islands near to S. America to those proper to the Continent. This struck me profoundly, especially the difference of the species in the adjoining islets in the Galapagos Archipelago. 3^{rdly} the relation of the living Edentata & Rodentia to the extinct species. I shall never forget my astonishment when I dug out a gigantic piece of armour like that of the living Armadillo.

Reflecting on these facts & collecting analogous ones, it seemed to me probable that allied species were descended from a common parent. But for some years I could not conceive how each form became so excellently adapted to its habits of life. I then began systematically to study domestic productions, & after a time saw clearly that man's selective power was the most important agent. I was prepared from having studied the habits of animals to appreciate the struggle for existence, & my work in Geology gave me some idea of the lapse of past time. Therefore when I happened to read "Malthus on population" the idea of Natural selection flashed on me. Of all the minor points, the last which I appreciated was the importance & cause of the principle of Divergence. I hope I have not wearied you with this little history of the "Origin"— . . .

. . . This letter was begun several weeks ago, but I have delayed finishing it from having little strength & other things to do. Will you have the kindness to tell this to Prof. Gegenbaur as an apology for not having thanked him for the honour he has done me in sending me his work. By a strange chance I dissected several months ago the hind foot of a toad & was particularly curious to understand what the additional bones were, & this point I see will now be explained to me. As I know from one of the papers which you have sent me that you have attended to Entomostraca it has occurred to me that you might like to have a copy of my Vol. on the Balanidæ, of which I have a spare copy & would with pleasure send it if you wish for it, & will tell me how to forward it.[1]

With sincere respect | Believe me my dear Sir | yours very faithfully | Charles Darwin

[1] *Living Cirripedia* (1854) or *Fossil Cirripedia* (1854).

[In 1864, CD was nominated to receive the Copley Medal of the Royal Society of London by George Busk, seconded by Hugh Falconer; it was the third year he had been nominated. The nomination was controversial, since the award of the medal to CD would suggest to many that the Royal Society endorsed the theory of natural selection. William Sharpey was the secretary of the Royal Society.]

From Hugh Falconer to William Sharpey   25 October 1864

Dép. Tarn [et] Garonne | Montauban
25th Oct. 1864.

My dear Sharpey

Busk and myself have made every effort to be back in London by the 27th. inst. but we have been persecuted by mishaps—through the breakdown of trains, diligences, &c., so that we have been sadly put out in our reckoning—and have lost some of the main objects that brought us round by this part of France—none of which were idle or unimportant.

Busk started yesterday for Paris from Briunquel to make sure of being present at the meeting of the R. Council on Thursday. He will tell you that there were strong reasons for my remaining behind him. But as I seconded the proposal of M^r. Darwin for the Copley Medal, in default of my presence at the first meeting, I beg that you will express my great regrets to the President and Council at not being there— and that I am very reluctantly detained. I shall certainly be in London (D.V.) by the second meeting on the 3^d prox^o. Meanwhile I solicit the favour of being heard, through you, respecting the grounds upon which I seconded Mr. Darwin's nomination for the Copley Medal.

Referring to the classified list—which I drew up—of Mr. Darwin's scientific labours, ranging through the wide field of 1. Geology; 2. Physical Geography; 3 Zoology; 4. Physiological Botany; 5 Genetic Biology; and to the power with which he has investigated whatever subject he has taken up, "Nullum tetegit quod non ornavit"[1] I am of opinion that Mr. Darwin is not only one of the most eminent naturalists of his day, but that hereafter he will be regarded as one of the *Great Naturalists of all Countries and of all time.* His early work on the structure and distribution of Coral reefs constitutes an era in the investigation of the subject. As a monographic labour it may be compared with Dr. Wells Essay upon Dew, as original, exhaustive and complete—containing the closest observation with large and important generalizations.

Among the Zoologists, his monographs upon the *Balanidæ* and *Lepadidæ*, Fossil and Recent in the Palæontographical and Ray Societies' Publications are held to be models of their kind.

In Physiological Botany, his recent researches, upon the Dimorphism of the Genital organs in certain plants, embodied in his papers in the Linnean Journal, on *Primula Linum* and *Lythrum*, are of the highest order of importance. They open a new mine of observation, upon a field which had been barely struck upon before. The same remark applies to his researches on the structure and various adaptations of the *Orchideous* flower, to a definite object connected with impregnation of the plants through the agency of insects with foreign pollen. There has not yet been time for their due influence being felt in the advancement of the science. But in either subject, they constitute [an] advance *per saltum.* I need not dwell upon the value of his Geological Researches, which won for him one of the earlier awards of the "Wollaston Medal"—from the Geological Society—the best of judges on the point.

And lastly Mr. Darwin's great essay on the Origin of Species by Natural Selection. This solemn and mysterious subject had been either so lightly or so grotesquely treated before, that it was hardly regarded as being within the bounds of legitimate philosophical investigation. Mr. Darwin after 20 years of the closest study and research, published his views, and it is sufficient to say that they instantly fixed the attention of mankind throughout the civilized world. That the efforts of a single mind should have arrived at success on a subject of such vast scope, and encompassed with such difficulties was more than could have been reasonably expected—and I am far from thinking that Charles Darwin has made out all his case. But he has treated it with such power and in such a philosophical and truth-seeking spirit and illustrated it with such a vast amount of original and collated observation, as fairly to have brought the subject within the bounds of rational scientific research. I consider this great essay on *Genetic Biology* to constitute a strong additional claim on behalf of Mr. Darwin for the Copley Medal.

In forming an estimate of the value and extent of Mr. Darwin's researches, due regard ought to be had to the circumstances under which they have been carried out—a pressure of unremitting disease—which has latterly left him not more than one or two hours of the day which he could call his own.

Yours sincerely | H. Falconer

[1] He touched nothing that he did not adorn.

From Hugh Falconer 3 November 186[4]

3 Novr. 186⟨4⟩ | 9½ p.m.

Private

My dear Darwin

My most hearty congratulations to yourself, Mrs Darwin & all to whom you are dear at Down. The Council R.S have awarded to you the Copley Medal, and never was it better bestowed. Council just up.

Your friends—including myself did not fail to stand up for "the Origin of Specs"—as establishing a strong claim.

Don't charge me with inconsistency—or fancy for a moment that I am a ⟨conv⟩ert! I think the work has rare—very rare merits—voila tout!

With kind regards to M^rs. Darwin

Yours Ever affec^ly | H. Falconer

Majority of 12: rest comb^d had 6

P.S. I returned last night from Spain via France. On Monday I was at Dijon where—while in the Museum—M. Brullé Professor of Zoology, asked me what was my frank opinion of Charles Darwins Doctrine? He told me in despair, that he could not get his pupils to listen to any thing from him except à la Darwin! The poor man, could not comprehend it—and was still unconvinced—but that all young Frenchmen would hear or believe nothing else.

H. F.

To T. H. Huxley 5 November [1864]

Down. | Bromley. | Kent. S.E.
Nov. 5th

My dear Huxley

I must & will answer you, for it is a real pleasure to me to thank you cordially for your note. Such notes, as this of yours & a few others are the real medal to me & not the round bit of gold. These have given me a pleasure which will long endure; so believe in my cordial thanks for your note.

I want to make a suggestion to you, but which may probably have occurred to you. Emma was reading your Lecture[1] to Horace & ended by saying "I wish he would write a book" I answered he has just written a great Book on the Skull.[2] "I dont call that a Book" she replied & added "I want something that people can read; he does write so well". Now with your ease in writing & with knowledge at your fingers' ends, do you not think you could write a "Popular Treatise on Zoology". Of course it would be some waste of time; but I have been asked more than a dozen times to recommend something for a beginner, & could only think of Carpenter's Zoology.[3] I am sure that a striking Treatise would do real service to Science by educating naturalists. If you were to keep a portfolio open for a couple of years, & throw in slips of paper, as subjects crossed your mind; you would soon have a skeleton (& that seems to me the difficulty) on which to put the flesh & colours in your inimitable manner.

I believe such a Book might have a brilliant success: but I did not intend to scribble so much about it.—

Give my kindest remembrances to Mrs Huxley & cordial thanks for her sympathy. Tell her I was looking at "Enoch Arden" & as I know how she admires Tennyson I must call her attention to two sweetly pretty lines (p. 105) . . . . : "and he meant, he said he meant, Perhaps he meant, or partly meant, you well".[4] Such a gem as this is enough to make me young again & like poetry with pristine fervour.—

My dear Huxley | Yours affecy | Ch. Darwin

Can you give me a Photographic Carte of yourself— I have set up a Book for my Scientific friends, | C. D.

1 CD may be referring to one of the lectures from the series *On our knowledge of the causes of the phenomena of organic nature* (London, 1862).

[2] T. H. Huxley, *Lectures on the elements of comparative anatomy. On the classification of animals and on the vertebrate skull* (London, 1864).
[3] W. B. Carpenter, *Zoology; being a systematic account of the general structure, habits, instincts, and uses of the principal families of the animal kingdom; as well as of the chief forms of fossil remains* (London, 1845).
[4] CD quotes from 'Sea dreams', by Alfred Tennyson. The poem appears in *Enoch Arden, etc* (London, 1864).

[In the preface to *Methods of study in natural history* (Boston, Mass., 1863), Louis Agassiz, professor of natural history at Harvard and father of Alexander Agassiz, stated his opposition to 'the transmutation theory', while allowing that others might hold contrary views 'in a spirit quite as reverential' as his own. He insisted that the theory was 'opposed to the processes of Nature, ... contradicted by the facts of Embryology and Paleontology', and that the experiments upon domesticated animals and cultivated plants, on which its adherents based their views, were 'entirely foreign to the matter in hand'.]

From B. D. Walsh 7 November 1864

Rock Island, Illinois, U.S.
Nov. 7. 1864

Chas. Darwin Esq.
Dear Sir,

Your welcome letter of Oct. 21 was received two days ago. I transmit you by this Mail a Paper of mine on "Certain Entomological Speculations of the New England Naturalists", in which I have taken occasion to show the absurdity of Agassiz's refutation of the "Origin of Species". Since it was published, I have received a long letter of seven pages of letter-paper from Alex. Agassiz, criticizing in a friendly way sundry passages in it. Amongst other things, he observes in regard to my charge that Prof. Agassiz had misstated your theory, that "Darwin in a recent letter to father *thanks him* for the manner in which he has conducted his opposition to his theory. So you see that opinions may differ on the misstatement of Darwin's Theory". In reply, I suggested that that fact merely showed that having been abused as an atheist, deist, infidel &c by other writers, you probably felt grateful to any writer who was willing to allow you "a spirit quite as reverential as his own". (*Meth. Study* Pref. p. iv.) ...

I do not know what the European Scientific World thinks of Agassiz, but here he is popularly considered as the incarnation of Science & as infallible as the Bible. He strikes me as a very much overrated man, who perpetually allows his imagination to get the better of his judgement, & who not unfrequently argues for victory instead of for truth. I have heard him myself in a Public Lecture assert in the most unqualified manner as a notorious truth that **all** the animals of every Geological Epoch were specifically distinct from those of the Epochs preceding & following it & that in the Eocene, Miocene &c periods there merely began to be some approximation to the present order of things. I am told by a New England correspondent, who belongs to his School, that he positively denies the truth of the facts upon which the very names Eocene &c are based. Do any European authorities uphold him in these views? I am told Pictet does.

You ask, "What can be the meaning or use of the great diversity of the external generative organs" in many groups of Insects? . . .

. . . there is a great diversity in the mode in which the *coitus* of insects is effected, & in almost every case the structure of the parts is such that the ♀ can, *if she chooses*, refuse coition. Hence I conceive has arisen the great variation in the structure & armature of the ♂ organs.

You express a desire to know what my life in this new country is. I will, at the risk of *boring* you, give you the particulars. When I left England in 1838, I was possessed with an absurd notion that I would live a perfectly natural life, independent of the whole world—*in meipso totus teres atque rotundus.*[1] So I bought several hundred acres of wild land in the wilderness, 20 miles from any settlement that you would call even a village, & with only a single neighbor. There I gradually opened a farm, working myself like a horse day after day, & raising great quantities of hogs & bullocks. Being far away from any mechanics, & having the natural gift of turning my hand to almost anything except blacksmithing, I did all kinds of jobs for myself, from mending a pair of boots to hooping a barrel or putting a new head into it, for the simple reason that it was less trouble to do such work myself than to hitch up my team & drive 10 or 20 miles to a mechanic's. I suppose I should at this present moment be vegetating there like a cabbage, but there was a Swedish Colony that settled in the neighborhood & dammed up the river everywhere for mill-seats, so that a great deal of malaria was generated & I had some severe attacks of fever, one of which brought me almost to death's door. So at the end of 12 years I sold out at a great sacrifice, & found that I was just $1000 poorer than when I began. I then moved to Rock Island & went into the Lumber business—"lumber" in America means all kinds of sawed boards & planks, "timber" being stuff that is hewed with the broad axe—& in seven years time had made a good many thousand dollars. I then, thinking I had been a slave about long enough, quitted that business, & invested most of my funds in building ten two-story brick houses for rent, upon the proceeds of which & a few of Uncle Sam's Bonds, I manage to live & keep all the time out of debt. If the times were what they used to be, I should have my whole time to myself & as large an income as I care about. As it is, what with the depreciation of such property & what with the war-taxes &c, I have to devote about $\frac{1}{4}$ of my time to executing sundry repairs on my tenements, & the remaining $\frac{3}{4}$ I devote to Entomology. I suppose I have the largest collection of Insects in America, except one or two in the great Atlantic Cities.

Will you favour me with your Photograph, & if convenient Prof. Westwood's, for my Entomological Album? I enclose you my own.

Believe me, my dear Sir, | very sincerely yours | Benj. D. Walsh

[1] 'In myself a whole, smoothed and rounded'. The phrase is adapted from Horace's *Satires*, 2.7.86.

From J. D. Hooker 2 December 1864

Kew
Dec 2d/64

Dear Darwin ...

Have you heard of the small breeze at R.S. apropos of your award— Busk told me—thus— Sabine said, in his address, that in awarding you the Copley, "all consideration of your Origin was expressly excluded"— After the address Huxley gets up & asks how this is—& being assured it is so, he insists on the minutes of Council being produced & read, in which of course there was no such exclusion or indeed any allusion to the Origin— Busk & Sabine afterwards were discussing the point—Sabine saying that no allusion = express exclusion, & shuffling as usual—when up comes Falconer & to Busk's horror compliments Sabines address unreservedly.— Busk thinking that F. had overheard the discussion said nothing at the time, but calls Falconer to account afterwards; upon which F. is grievously put out, at finding out what he has done, & forthwith goes & writes a letter to Sabine on the subject— May the Lord have mercy on S. is all I can say; for F. will have none

This is the story as I believe Busk to have told it me yesterday: but as it has thus passed through 2 hands I do not doubt it is damaged in the process,—so pray take it for no more than it is worth—

Wistaria certainly twined right up a *Salisburia* upwards of 6 in diam in our Garden. & I think I may positively affirm that *Ruscus androgynus* is twining sua sponte,[1] round one of the columns of our new winter Garden. which is 9 in. diam. . . .

Ever yrs affec | J D Hooker

[1] Of its own accord.

[Hooker enclosed this letter with his letter to CD of [6 December 1864].]

From T. H. Huxley to J. D. Hooker 3 December 1864

Jermyn S.
Dc^{r} 3rd 1864

My dear Hooker . . .

I wish you had been at the Anniversary Meeting & Dinner, because the latter was very pleasant & the former, to me, very disagreeable My distrust of Sabine is as you know, chronic—: and I went determined to keep careful watch on his address—lest some crafty phrase injurious to Darwin should be introduced— My suspicions were justified— The only part of the address to Darwin written by Sabine himself containing the following passage

"Speaking generally and collectively we have expressly omitted it (Darwins theory) from the grounds of our award"

Of course this would be interpreted by every body as meaning that after due discussion the council had formally resolved not only to exclude Darwins theory from the grounds of the award but to give public notice through the President that they had done so—and furthermore that Darwins friends had been base enough to accept an honour for him on the understanding that in receiving it he should be publicly insulted—!

I felt that this would never do, and therefore when the resolution for printing the address was moved—I made a speech which I took care to keep perfectly cool & temperate, disavowing all intention of interfering with the liberty of the President to say what he pleased—but exercising my constitutional right of requiring the minutes of council making the award to be read—in order that the Society might be informed whether the conditions implied by Sabine had been imposed or not—

The resolution was read & of course nothing of the kind appeared

Sabine didn't exactly like it I believe Both Busk & Falconer protested against the passage to him—and I hope it will be withdrawn when the address is printed[1]

If not there will be an awful row—and I for one will shew that old fox no mercy.

Ever yours faithfully | T. H. Huxley

[1] In the version of Sabine's address that was eventually published in the Royal Society's *Proceedings*, the wording of the controversial passage relating to *Origin* was changed from 'we have expressly omitted it from the grounds of our award' to 'we have not included it in our award'. For a comparison of the different versions, see *Correspondence* vol. 12, Appendix IV.

To B. D. Walsh 4 December [1864]

Down. | Bromley. | Kent. S.E.
Dec. 4th.—

My dear Sir

I have been greatly interested by your account of your American life. What an extraordinary & self-contained life you have led! & what vigour of mind you must possess to follow science with so much ardour after all that you have undergone.— I am very much obliged for your pamphlet on Geograph. Distrib,—on Agassiz &c.— I am delighted at the manner in which you have bearded this lion in his den. I agree most entirely with all that you have written. What I meant, when I wrote to Agassiz to thank him for a bundle of his publications, was exactly what you suppose.

I confess, however, I did not *fully* perceive how he had mistated my views; but I only skimmed through his "Method of Study" & thought it a very poor Book.— I am so much accustomed to be utterly misrepresented that it hardly excites my attention. But you really have hit the nail on the head *capitally*. All the younger good naturalists, whom I know think of Agassiz as you do; but he did grand service about Glaciers & Fish.— About the succession of forms, Pictet has given up his whole views, & *no* geologist now agrees with Agassiz.— I am glad that you have attacked Dana's wild notions: I have a great respect for Dana, but I declare I fear that his long illness has somewhat enfeebled his brain.— If you have opportunity read in Transact. Linn. Soc.[y] Bates on mimetic Lepidoptera of Amazons; I was delighted with his paper.— . . .

Many thanks about the connexion of male & female insects & their organs.— It occurred to me as just possible that the organs for oviposition might be very different in allied species; & that this might lead by correlation to differences in the male parts; but this was a simple groundless conjecture on my part, & not applicable anyhow to Bombus. Would species of Bombus copulate differently? . . .

As you allude in your paper to the believers in change of species, you will be glad to hear that very many of the very best men are coming round in Germany.— . . . So it is, I hear, with the *younger* Frenchmen.—

Pray believe me | My dear Sir | Yours very sincerely | Ch. Darwin

1865

From J. D. Hooker 1 January 1865

Kew
Jany 1/65

Dear Darwin ...

I have read Sabines complete address (I had seen only extracts before) & am indignant & disgusted at the mutilation & emasculation of what I wrote— Especially about *Lythrum* & *Linum*, which he has made nonsense of & the use your observations will be in interpreting, no end of phenomena not yet guessed at. Poor old man, he is ill still, & I am beginning to fear that my ill-natured prophecy, that the Presidentship would be the death of him, may come true.—

Have you read Huxleys (I suppose) slashing leader in todays Reader.[1] it is uncommonly able &c: but as usual with him, he goes like a desert whirlwind over the ground scorching blasting & suffocating all opposing objects, & leaving nothing but dry bones on the ground. The vegetation he withers was one of vile weeds to be sure, but vile weeds are *green*, & all is *black* after him ...

My book on Geog. Distrib. is *nowhere*— I wish it were only begun. . . .

Ever y^{rs} affec | J. D. Hooker.

[1] Thomas Henry Huxley's unsigned article 'Science and "Church policy"', which appeared in the 31 December 1864 issue of the *Reader*, p. 821, criticised 'leading statesmen' and 'ecclesiastical dignitaries' for their lack of regard for science, and addressed in particular the remarks on science made by Benjamin Disraeli in a recent speech on church policy.

From Henrietta Anne Huxley 1 January 1865

Dear M^r. Darwin

Hal has just brought me your note containing your slyly disparaging remarks on my beloved Tennyson—& quoting "as a gem"

'And he meant, he said
he meant,
Perhaps he meant, or partly
meant you well.'

In the first place it was very mean of you to give the lines without the context shockingly Owenlike

Secondly. The lines only convince me more than ever that Tennyson is quite master of his situation. Could you better render In words, the desire in the wife's mind to do justice, to—her enemy I suppose for I have not read "Sea Dreams", together with the conflicting feeling which yet possessed her of his insincerity? I am very pleased that Tennyson accredits the feminine mind with such a strong sense of justice.

I now refer to the book— I am grieved to find that a philosopher of your repute—should have damaged your reputation for accuracy so greatly as to tell me that the quotation was from "Enoch Arden" whereas it was from "Sea Dreams"— If the "facts?!" in the Origin of Species are of this sort—I agree with the Bishop of Oxford—

Yours too sincerely | Henrietta Huxley

love to your dear wife & ask her for a screed.

To T. H. Huxley 4 January [1865]

Down. | *Bromley.* | *Kent. S.E.*
Jan. 4th

My dear Huxley

Very many thanks for your Photograph, which is excellent, but it makes you look too black & solemn as if facing the bench of Bishops.—

We were all charmed with Mrs. Huxley "too sincere" note. Oh that I should live to be called "Owen-like"! I was indeed innocent of concealing the context, for I did not read one line beyond the charming lines which I quoted, & they were enough for me!

How hard you are worked & I do wish that you had more leisure or at least not so many lectures. It is an absolute marvel to me how much you do.— I knew there was very little chance of your having time to write a popular treatise on Zoology; but you are about the one man that could do it. At the time I felt it would be almost a sin for you to do it, as it would of course destroy some original work. On the other hand, I sometimes think that general & popular Treatises are almost as important for the progress of science as original work.— As for writing being a great labour to you, I can hardly swallow that. Your words on paper seem always to come out spontaneously. . . .

I am no great thing in health, but manage most days to do a little work.—

Our kindest remembrances to Mrs Huxley | Ever yours
C. Darwin

From Charles Lyell 16 January 1865

Magdeburg:
January 16, 1865.

My dear Darwin,—

I was so busy with the last chapters of my new edition of the 'Elements'[1] before I left town a month ago, that I did not reply to your kind letter about my after-dinner speech on your Copley medal at the Royal Society anniversary. I have some notes of it, and hope one day to run over it with you, especially as it was somewhat of a confession of faith as to the 'Origin.' I said I had been forced to give up my old faith without thoroughly seeing my way to a new one. But I think you would have been satisfied with the length I went. The Duke of Argyll expresses in his address to the Edinburgh Royal Society[2] very much what I have done ('Antiquity of Man,' p. 469),[3] that variation or natural selection cannot be confounded with the creational law without such a deification of them as exaggerates their influence. He seems to me to have put the difficulty pretty clearly, but on the other hand he has not brought out as fully as I should have liked him to have done, the great body of evidence so admirably brought to bear in your work, in proof of the bond of mutual descent, and the manner in which species and genera branched from common ancestors. He did not entertain this idea till he had read your book, and he is now evidently impressed with it, as I am; and he would, I think, go the whole length, were it not for the necessity of admitting, in order to be consistent, that man and the quadrumana came from a common stock. He does, indeed, in defiance of consistency, admit for the humming-birds what he will not admit for the *primates*, and Guizot's theology is introduced to support him;[4] but the address is a great step towards your views—far greater, I believe, than it seems when read merely with reference to criticisms and objections. The reasoning about materialism appears to me admirably put, and his definition of the various senses in which we use the term 'law'; though, having only read the speech once, I am not yet able to judge critically on all these points. He assumes far too confidently that the colours of the humming-birds are for mere ornament and beauty. I can conceive a meaning in your sense for the advantage of the creature, or of its friends and enemies, in every coloured ray of light reflected from the plumes. We must indeed know far more than we do before we can dogmatise on the irrelevancy of particular colours to the well-being of a species. He ought also to define beauty, and tell us whether it is in reference to man or bird. I have no objection to the idea of beauty or variety for its own sake, but to assume it so positively is unphilosophical.

We have been about three weeks at Berlin, and I had . . . an animated conversation on Darwinism with the Princess Royal, who is a worthy daughter of her father, in the reading of good books and thinking of what she reads.[5] She was very much *au fait* at the 'Origin' and Huxley's book,[6] the 'Antiquity,' &c. &c., and with the Pfahlbauten [lake-dwelling] Museums which she lately saw in Switzerland. She said after twice reading you she could not see her way as to the origin of four things; namely the world, species, man, or the black and white races. Did one of the latter come from the other, or both from some common stock? And she asked me what I was doing, and I explained that in recasting the 'Principles'[7] I had to give up the independent creation of each species. She said she fully understood my difficulty, for after your book 'the old opinions had received a shake from which they never would recover.' I shall be very glad to hear what you think of the Duke of Argyll's comments on the 'Origin'. I think that your book is a vast step towards showing the methods which have been followed in creation, which is as much as science can ever reach, and the Duke, I think, has not fully appreciated the advance which has been made, even in his own mind.

I had hoped that a copy of the 'Elements' would have been sent to you while I was still at Berlin. You will find much that is new, and nothing, I think, clashing with the 'Origin'. Please read my description of the Atlantis theory. I fear I shall return and find the book still unborn, which is too bad of the printer. Please let me know how your health has been during the last four weeks.

Ever most truly yours | Charles Lyell.

[1] Charles Lyell, *Elements of geology*, 6th edition (London, 1865).
[2] G. D. Campbell, eighth duke of Argyll, 'Opening address, 1864–5 session', *Proceedings of the Royal Society of Edinburgh* 5 (1862–6): 264–92.
[3] Charles Lyell, *The geological evidences of the antiquity of man*, 3d edition (London, 1863).
[4] François Guizot maintained that humans could not have survived unless they had appeared with all of their faculties and powers fully developed.
[5] Victoria Adelaide Mary Louise, the eldest child of Queen Victoria and Prince Albert, was married to the crown prince of Prussia.
[6] T. H. Huxley, *Evidence as to man's place in nature* (London, 1863).
[7] Charles Lyell, *Principles of geology*, 9th edition (London, 1853).

To Charles Lyell 22 January [1865]

Down. | Bromley. | Kent. S.E.
Jan 22

My dear Lyell

I thank you for your very interesting letter. I have the true English instinctive reverence for rank & therefore liked to hear about the Princess Royal. You ask what I think of the Duke's address & I shall be glad to tell you.

It seems to me *extremely* clever like every thing that I have read of his; but I am not shaken; perhaps you will say that neither gods nor men could shake me. I demur to the Duke reiterating his objection that the brilliant plumage of the male humming bird could not have been acquired through selection, at the same time entirely ignoring my discussion (p. 93 3rd Edition) on beautiful plumage being acquired thro' *sexual* selection. The Duke may think this insufficient, but that is another question. All analogy makes me quite disagree with the Duke that the differences in the beak, wing & tail are not of importance to the several species. In the only two species which I have watched, the difference in flight & in the use of the tail was conspicuously great.

The Duke who knows my orchis book so well might have learnt a lesson of caution from it, with respect to his doctrine of differences for mere variety or beauty. It may be confidently said that no tribe of plants presents such grotesque & beautiful differences which no one until lately conjectured were of any use; but now in almost every case, I have been able to shew their important service.

It should be remembered that with humming birds or orchids a modification in one part will cause correlated changes in other parts. I agree with what you say about beauty. I formerly thought a good deal on the subject & was led quite to repudiate the doctrine of beauty being created for beauty's sake. . . .

The more I work the more I feel convinced that it is by the accumulation of such extremely slight variations that new species arise. I do not plead guilty to the Duke's charge that I forget that natural selection means only the preservation of variations which independently arise. I have expressed this in as strong language as I could use; but it w^{d} have been infinitely tedious had I on every occasion thus guarded myself. I will cry "peccavi"[1] when I hear of the Duke or you attacking Breeders for saying that man has made his improved Shorthorns or Pouter-pigeons or Bantams. And I c^{d} quote still stronger expressions used by agricuturalists. Man does make his artificial breeds, for his selective power is of such importance relatively to that of the slight spontaneous variations. But no one will attack Breeders for using such expressions, & the rising generation will not blame me.

Many thanks for your offer of sending me the Elements; I hope to read it all, but unfortunately reading makes my head whiz more than any thing else. I am able most days to work for 2 or 3 hours & this makes all the difference in my happiness. I have resolved not to be tempted astray, & to publish nothing till my Vol. on Variation is completed.

We (dictater & writer)[2] send our best love to Lady Lyell | yours affectionately | Charles Darwin

If ever you sh^{d}. speak with the Duke on the subject please say how much interested I was with his Address & tell him about Sexual Selection.

[1] I have sinned.

[2] The letter is in Emma Darwin's hand.

[Robert FitzRoy, captain of HMS *Beagle* during CD's 1831 to 1836 voyage, committed suicide on 30 April 1865.]

To J. D. Hooker 4 May [1865]

Down
May 4th

My dear Hooker

Sincere thanks for two most interesting notes. I was astounded at news about FitzRoy; but I ought not to have been, for I remember once thinking it likely; poor fellow his mind was quite out of balance once during our voyage. I never knew in my life so mixed a character. Always much to love & I once loved him sincerely; but so bad a temper & so given to take offence, that I gradually quite lost my love & wished only to keep out of contact with him. Twice he quarrelled bitterly with me, without any just provocation on my part. But certainly there was much noble & exalted in his character. Poor fellow his career is sadly closed. You know he was nephew to L[d]. Castlereagh, & very like him in manners & appearance.—

You will be glad to hear, I know my dear old friend, that my ten or 12 days of sickness has suddenly ceased, & has left me not much the worse: I feared it was the beginning of another six or nine months miserable attack & feared it much: Jenner has been here, & is evidently perplexed at my case; he struck me as a more able & sensible man, than he did before, for then I could not talk with him: I shall consult no one else.— . . .

I do not suppose you will care to hear, apropos to your Rafflesia but in Siebold's "Parthenogenesis" of Bee p. 107, there are cases of the 2 sexes of the same Gall-insect being produced from different plants.—[1]

I have just received a paper from Häckel, which gives an astonishing case of propagation in a Medusa; it is exactly as if tadpoles were of two sexes & regularly laid eggs, but also produced by *budding* frogs, these likewise being of 2 sexes & laying eggs!

Very many thanks to you for writing to me about FitzRoy.— Poor fellow how kind he was to me at first during the voyage.—
. . .

Yours ever affectionately | C. Darwin . . .

[1] K. T. E. von Siebold, *On a true parthenogenesis in moths and bees; a contribution to the history of reproduction in animals*. Translated by W. S. Dallas. (London, 1857).

[With the fall of Richmond, Virginia, the capital of the Confederacy, on 2 and 3 April, and the surrender of Robert E. Lee's Army of Northern Virginia on 9 April 1865, the American Civil War had effectively come to an end.]

From Asa Gray 15 and 17 May 1865

Cambridge, Mass.
May 15, 1865—

My Dear Darwin

Your kind letter of the 19th ult. crossed a brief note from me. I am too much *distracted* with work at this season to write letters on our affairs, and if I once begin, I should not know where to stop. You have always been sympathising and just, and I appreciate your hearty congratulations on the success of our just endeavors. You have since had much more to rejoice over, as well as to sorrow with us. But the noble manner in which our country has borne itself should give you real satisfaction. We appreciate too the good feeling of England in its hearty grief at the murder of Lincoln.

Don't talk about our "hating" you,—nor suppose that we want to rob you of *Canada*—for which nobody cares.

We think we have been ill-used by you, when you thought us weak and broken.— & when we expected better things. We have learned that we *must be strong* to live in peace & comfort with England,—otherwise we should have to eat much dirt. But now that we are on our feet again, all will go well, and *hatred* will disappear. Indeed, I see little of that. We do not even hate the Rebels, and may not even execute so much of justice as to convict of treason & hang their President, whom we have just caught,—but *I hope we shall*,—hang the leader & spare the subordinates. We are now feeding the south, who starved our men taken prisoners.

Slavery is thoroughly dead. We have a deal to do, but shall do it, I trust, and deserve your continued approbation. We have a load to carry—heavy, no doubt, but a young & re-invigorated country, with a future before it can do and bear, & prosper under what might stagger a full-grown, mature country of the Old world

I must look to the *Plantago* dimorphism: for, as you say, these plants, fertilised by wind, could gain nothing by being dimorphic. No dimorphic species grows very near here,—nor can I now get seeds of P. Virginica. Perhaps a good look at even dried specimens, under your hints, may settle the matter.

I was exceedingly interested with the Lythrum paper (but had no time to write a notice of it.), & I wait expectingly for your Climbing plants. You are the very prince of investigators. . . .

In great haste, dear Darwin, | Your affectionate | A. Gray . . .

Thanks for the Times.—apparently from you.

[By May 1865, CD had formulated an ambitious hypothesis, intended to offer a comprehensive account of heredity and development. He believed the hypothesis, which he came to call 'pangenesis', could explain both sexual and asexual reproduction, as well as reversion and the regrowth of body parts. CD suggested that each individual cell threw off minute particles (gemmules) that circulated in the bodily fluids and were capable of generating new cells, or remaining dormant until required. The gemmules were drawn together by mutual attraction and into the right places in the right order to, for instance, reconstruct (in organisms that had that capability) a tail or other limb where one had been torn off. Cells threw off gemmules at all stages of development and all gemmules circulated until required: thus embryo gemmules thrown off by embryo cells came together, when conditions were right, to form what CD called the germ (the female element) or the male element; the germ, if fertilised (or even if not, in the case of parthenogenesis), grew into a new embryo. Gemmules could also remain dormant for several generations before being sparked into activity.]

To T. H. Huxley 27 May [1865]

Down. | *Bromley.* | *Kent. S.E.*
May 27th.—

My dear Huxley . . .

I write now to ask a favour of you, a very great favour from one so hard worked as you are. It is to read 30 pages of M.S, excellently copied out, & give me not lengthened criticism, but your opinion whether I may venture to publish it. You may keep the M.S. for a month or two. I wd not ask this favour, but I *really* know no one else whose judgment on the subject wd be final with me.

The case stands thus; in my next book I shall publish long chapters on bud—& seminal—variation, on inheritance, reversion, effects of use & disuse &c. I have also for many years speculated on the different forms of reproduction. Hence it has come to be a passion with me to try to connect all such facts by some sort of hypothesis. The M.S which I wish to send you gives such a hypothesis; it is a very rash & crude hypothesis yet it has been a considerable relief to my mind, & I can hang on it a good many groups of facts. I well know that a mere hypothesis, & this is nothing more, is of little value; but it is very useful to me as serving as a kind of summary for certain chapters. Now I earnestly wish for your verdict given briefly as "Burn it"—or, which is the most favorable verdict that I can hope for, "It does rudely connect together certain facts & I do not think it will immediately pass out of my mind" If you can say this much & you do not think it absolutely ridiculous I shall publish it in my concluding Chapter. Now will you grant me this favour? You must refuse if you are too much over worked—

I must say for myself that I am a hero to expose my hypothesis to the fiery ordeal of your criticism

Believe me my dear Huxley | yours most sincerely | Ch. Darwin

From T. H. Huxley 1 June 1865

Museum of Practical Geology
June 1 | 1865

My dear Darwin

Your M.S reached me safely last evening—

I could not refrain from glancing over it on the spot and I perceive I shall have to put on my sharpest spectacles & best considering cap—

I shall not write till I have thought well on the whole subject

Ever Yours | T H Huxley

To T. H. Huxley 12 July [1865]

Down Bromley | Kent
July 12

My dear Huxley

I thank you most sincerely for having so carefully considered my M.S.[1] It has been a real act of kindness. It w^{d} have annoyed me extremely to have republished Buffon's views, which I did not know of but I will get the book; & if I have strength I will also read Bonnet. I do not doubt your judgment is perfectly just & I will try to persuade myself not to publish. The whole affair is certainly much too speculative; yet I think some such view will have to be adopted, when I call to mind such facts as the inherited effects of use & disuse &c. But I will try to be cautious Any how I shall have plenty of time for consideration for my health has been so bad of late that I have written nothing during the last 2 months.

Again accept my sincere thanks & believe me my dear Huxley yours very truly | Ch. Darwin

you have been very good to take so much trouble.— . . .

[1] Huxley's letter has not been found.

From T. H. Huxley 16 July 1865

Jermyn St

July 16th | 1865

My dear Darwin

I have just counted the pages of your M.S. to see that they are all right and packed it up to send you by post, registered, so I hope it will reach you safely— I should have sent it yesterday but people came in & bothered me about post time

I did not at all mean by what I said to stop you from publishing your views and I really should not like to take that responsibility

Somebody rummaging among your papers half a century hence will find Pangenesis & say "See this wonderful anticipation of our modern Theories—and that stupid ass, Huxley, prevented his publishing them"

And then the Carlylians of that day will make me a text for holding forth upon the difference between mere vulpine sharpness & genius—[1]

I am not going to be made a horrid example of in that way— But all I say is publish your views—not so much in the shape of formed conclusions—: as of hypothetical developments of the only clue at present accessible—and don't give the Philistines more chances of blaspheming than you can help

I am very grieved to hear that you have been so ill again—
Ever | Yours faithfully | T. H. Huxley

[1] In *On heroes, hero-worship, & the heroic in history* (London, 1841), p. 173, Thomas Carlyle compared the virtuous man's ability to understand nature with that of the immoral man. The latter he likened to a fox and said: 'what such can know of Nature is mean, superficial, small'.

To T. H. Huxley [17 July 1865]

Down
Monday

My dear Huxley

Forgive my writing in pencil, as I can do so lying down— I have read Buffon:— whole pages are laughably like mine. It is surprising how candid it makes one to see one's views in another man's words— I am rather ashamed of the whole affair, but not converted to a no-belief— What a kindness you have done me with your "*vulpine sharpness*".—

Nevertheless there is a fundamental distinction between Buffon's views & mine. He does not suppose that each cell or atom of tissue throws off a little bud; but he supposes that the sap or blood includes his "organic molecules", *which are ready formed*, fit to nourish each organ, & when this is fully formed, they collect to form buds & the sexual elements:— It is all rubbish to speculate as I have done; yet, if I ever have strength to publish my next book, I fear I shall not resist "pangenenesis", but I assure you I will put it humbly enough— The ordinary course of development of beings, such as the echinodermata, in which new organs are formed at quite remote spots from the analogous previous parts, seems to me extremely difficult to reconcile on any view, except the free diffusion in the parent of the germs or gemmules of each separate new organ; & so in cases of alternate generation.— But I will not scribble any more. Hearty thanks to you, you best of critics & most learned man.

Yours most truly | C. Darwin …

Your last note made us all laugh.— The future rummager of my papers will I fear, make widely opposite remarks—

From Asa Gray 24 July 1865

Cambridge, Mass.
July 24, 1865.

My Dear Darwin

I had heard, thro' Hooker, that you had been poorly again, and I think that a letter, signifying my sorrow was written and has crossed yours just received. I fancy you as now getting much better again. I am glad you *did* not see Brace (tho' sorry you *could* not): he is a great *talker*, or rather *questioner*, and would have exhausted you terribly.

Here, at length are some seeds of *Specularia perfoliata*, from D[r]. Engelmann. I have sown half, and send you the rest. I am reading in snatches, your admirable paper on Climbing Plants—as yet only 88 pages of it,—and am watching, with great interest, all the climbers I have at hand. What a nice piece of work you have made of it!

I see you explain & illustrate at length the double turn of a caught tendril. Is it not enough to say that, with both ends fixed, if it shortens say by the contraction of one side it must, by mechanical necessity turn its coil different ways, from a neutral point.

I am vexed to have no *Adlumia* in the garden this year—i.e. no seedlings came up last year—from the drought, I suppose. I may meet with it, in the country—the Western part of New York—where my wife and I are going—two days hence—to rusticate for 3 weeks— She needs it.—tho' I hardly do. The first thing I do will be to write, for Sill. Journal[1] an analysis of your great paper, And when I return, if not too late, I mean to give 2 or 3 lectures from it to my University Class. . . .

Ere this Mrs. Wedgwood[2] should be back from Canada, but I have not learned that she is so. She was to let me know, and we would have a day on the shore where Mr. Loring[3] lives in summer—a pretty bit of country. But it is now too late.

I wish she could have been here on Friday, when we welcomed back our Harvard men who had been in the war, over 500 of them—and remembered those who had died for their country. What a day we had!

Jefferson Davis richly deserves to be hung. We are all willing to leave the case in the hands of the Government, who must take the responsibility. If I were responsible, I would have him tried for *treason*—the worst of crimes in a republic— convicted, sentenced to death,—and then I think I should commute the penalty, not out of any consideration for him, but from policy, and for his more complete humiliation. The only letters I have received expressing a desire to hang him, are from rebeldom itself—from Alabama. You see slavery is *dead*, **dead**,—an absolute unanimity as to this. The Revolted States will behave as badly as they can, but they are so thoroughly whipped that can't stir, hand or foot,—and we are disbanding all our armies— a corporal's guard is enough to hold South Carolina—

Seriously, there are difficult questions before us,—but only one result is possible— the South must be renovated, and Yankeefied.

Well— take good care of yourself, and let me know that you are again in comfortable condition

Ever Your affectionate friend | A. Gray

1 Silliman's Journal: *The American Journal of Science and Arts.*

2 Frances Wedgwood.

3 Charles Greely Loring was Gray's father-in-law.

From A. R. Wallace 18 September 1865

9, St. Mark's Crescent, | *Regent's Park,* N.W.
Septr. 18th. *1865*

Dear Darwin

I should have written before to thank you for the copy of your paper on climbing plants, which I read with great interest; I can imagine how much pleasure the working out must have given you. I was afraid you were too ill to make it advisable that you should be bothered with letters.

I write now in hopes you are better, to communicate a curious case of *variation* becoming at once *hereditary*, which was brought forward at the Brit. Ass^n^.

I send a note of it on the other side, but if you would like more exact particulars, with names & dates and a drawing of the bird, I am sure Mr. O'Callaghan would send them you.

I hope to hear that you are better & that your new book is really to come out next winter.

Believe me | Yours very faithfully | Alfred R. Wallace

C. Darwin Esq.

Note

Last spring Mr. O'Callaghan was told by a country boy that he had seen a blackbird with a topknot; on which Mr. O'C. very judiciously told him to watch it & communicate further with him. After a time the boy told him he had found a blackbird's nest & had seen this crested bird near it & believed he belonged to it— He continued watching the nest till the young were hatched. After a time he told Mr O'C. that two of the young birds seemed as if they would have topknots. He was told to get one of them as soon as it was fledged. However he was too late and they left the nest, but luckily he found them near & knocked one down with a stone, which Mr. O'C. had stuffed & exhibited. It has a fine crest, something like that of a Polish fowl but *larger* in proportion to the bird, & very regular & well formed. The male must have been almost like the Umbrella bird in miniature, the crest is so large & expanded.

ARW.

To A. R. Wallace 22 September [1865]

Down. | Bromley. | Kent. S.E.
Sep 22

Dear Wallace

I am much obliged for your extract; I never heard of such a case, though such a variation is perhaps the most likely of any to occur in a state of nature & be inherited, inasmuch as all domesticated birds present races with a tuft or with reversed feathers on their heads. I have sometimes thought that the progenitor of the whole class must have been a crested animal.

Do you make any progress with your journal of travels? I am the more anxious that you sh^{d} do so as I have lately read with much interest some papers by you on the Ouran Outang &c—in the Annals of which I have lately been reading the latter volumes. I have always thought that Journals of this nature do considerable good by advancing the taste for Natural history; I know in my own case that nothing ever stimulated my zeal so much as reading Humboldt's Personal Narrative.[1]

I have not yet rec^{d} the last part of Linn. Tran. but your paper at present will be rather beyond my strength, for tho' somewhat better I can as yet do hardly anything but lie on the sofa & be read aloud to.[2] By the way have you read Tyler & Lecky.[3] Both these books have interested me much. I suppose you have read Lubbock.[4] In the last Chap. there is a note about you in which I most cordially concur. I see you were at the Brit. Assoc. but I have heard nothing of it except what I have picked up in the Reader. I have heard a rumour that the Reader is sold to the Anthrop. Soc. If you do not begrudge the trouble of another note (for my sole channel of news thro' Hooker is closed by his illness) I sh[d] much like to hear whether the Reader is thus sold. I sh[d] be very sorry for it as the paper w[d] thus become sectional in its tendency. If you write tell me what you are doing yourself.

The only news which I have about the Origin is that Fritz Müller published a few months ago a remarkable book in its favour[5] & 2[ndly] that a 2nd French edition is just coming out

Believe me dear Wallace | yours very sincerely | Ch. Darwin

[1] Alexander von Humboldt, *Personal narrative of travels to the equinoctial regions of the New Continent, during the years 1799–1804*, translated into English by H. M. Williams, 7 vols. (London, 1814–29). On CD's enthusiasm for Humboldt, see also *Charles Darwin's letters: a selection 1825–1859*, edited by Frederick Burkhardt (Cambridge, 1996).

[2] A. R. Wallace, 'On the phenomena of variation and geographical distribution as illustrated by the Papilionidæ of the Malayan region', *Transactions of the Linnean Society of London* 25 (1865–6): 1–71.

[3] E. B. Tylor, *Researches into the early history of mankind* (London, 1865). W. E. H. Lecky, *History of the rise and influence of the spirit of rationalism in Europe* (London, 1865).

[4] John Lubbock, *Pre-historic times, as illustrated by ancient remains, and the manners and customs of modern savages* (London and Edinburgh, 1865).

[5] Fritz Müller, *Für Darwin* (Leipzig, 1864).

1866

To a local landowner [1866?][1]

Dear Sir.

As you are now so little on your Farm, you may not be aware that the necks of your horses are badly galled, as I have been informed by two persons. I hope you will immediately attend to this, for though I sh^{d} be very sorry to give trouble or annoyance to yourself from whom I have always received much civility, I must for the sake of humanity attend to this. A conviction for working Horses with galled necks is easily obtainable, on sufficient evidence being produced & I was most unwillingly compelled, after vainly remonstrating with Mr Ainslie by letter on the subject, to get the Officer of the Royal Humane Soc. to come down here & attend to the case, & Mr Ainslie was fined by the Magistrates at the Bromley Session.— I sincerely hope that you will at once make enquiries & give strict orders to your Bailiff not to work any horse with a wounded neck—

& | I remain | Dear Sir | Y^{rs} faithfully | Ch. Darwin

[1] This letter exists in the form of a draft only.

To Henry Bence Jones 3 January [1866]

Down. | Bromley. | Kent. S.E.
Jan 3.

My dear Dr Bence Jones

I have a good report to make. I am able now to walk daily on an average $3\frac{1}{2}$ miles & often one mile at a stretch.

My weight now with slight fluctuations keeps steady at the lowest point to which it has sunk. I feel altogether much more vigorous & active. I read more, & what is delightful, I am able to write easy work for about $1\frac{1}{2}$ hours every day. The only drawback is that on most days 3 hours after luncheon or dinner, I have a sharpish headache on one side, & with bad flatulence lasting to the next meal. I forgot to say that taking the whole day, the flatulence is somewhat diminished especially on my better days. One day when my head & stomach were extra bad, in despair I took a cup of coffee without sugar, & it acted really like a charm & has continued to do so; for I now take a cup of coffee each day with luncheon or dinner, & I believe I have never once had headache and flatulence after the meal with coffee. I have transposed luncheon & dinner & made other changes, but as far as I can discover it is the coffee which is effectual. Under these circumstances may I try coffee with both luncheon & dinner. I have not yet much taste for common meat, but eat a little game or fowl twice a day & eggs, omelet or maccaroni or cheese at the other meals & these I think suit me best. I have not taken to *[2 words illeg]* much starch for I have such horror about acid.

There is an odd change in my stomach, for the last 20 years coffee & cheese have disagreed with me, now they suit me eminently well. I took 10^{g} oxyde of Iron for a fortnight but did not miss it when I left it off 10 days ago: I will do as you like about retaking it. I have taken 10 drops of Muriatic acid twice a day (with Cayenne & ginger) for above 3 weeks & it suits me *excellently.* May I continue it longer? I hope you will be pleased with my report. I shall be grateful for any further advice

yours very faithfully with | cordial thanks | Ch. Darwin

[Emily Catherine Langton (Catherine) was CD's younger sister; she died at the end of January 1866.]

From Emily Catherine Langton to Emma and Charles Darwin
[6 and 7? January 1866]

Dearest Emma & Charles

I am so rapidly weaker I can lose no time in sending you all & Elizabeth my dearest farewell.[1] It is grievous to think I shall never see any of your dear faces. On New Year's day I knew this, and what a different world it seems to me.

What I want to say is that poor Susan feels my loss so cruelly—[2] I left off this last night as I was too exhausted to go on—

I am grieved indeed at poor Susan's loneliness, but there seems no help.

My dearest husband will feel my loss too; *what a nurse he is*, if he was not deaf—

Every body's love & goodness to me are past speech—

May God bless you all & may we meet hereafter.

E.C.L—

Sunday.

[1] Sarah Elizabeth Wedgwood, known as Elizabeth, was Emma Darwin's sister and Catherine's cousin.

[2] Susan Elizabeth Darwin, CD and Catherine's sister.

To A. R. Wallace 22 January 1866

Down. | Bromley. | Kent. S.E.
Jan 22. 1866

My dear Wallace

I thank you for your paper on Pigeons, which interested me, as every thing that you write does.[1] Who w[d] ever have dreamed that monkeys influenced the distribution of pigeons & parrots!

But I have had a still higher satisfaction; for I finished yesterday your paper in Linn. Trans.[2] It is admirably done. I cannot conceive that the most firm believer in Species c[d] read it without being staggered. Such papers will make many more converts among naturalists than long-winded books such as I shall write if I have strength.

I have been particularly struck with your remarks on Dimorphism; but I cannot quite understand one point (p. 22) & sh[d] be grateful for an explanation for I want fully to understand you.

How can one female form be selected & the intermediate forms die out, without also the other extreme form also dying out from not having the advantages of the first selected form; for as I understand, both female forms occur on the same Island. I quite agree with your distinction between dimorphic forms & varieties; but I doubt whether your criterion of dimorphic forms not producing intermediate offspring will suffice; for I know of a good many varieties which must be so called, that will not blend or intermix, but produce offspring quite like either parent.

I have been particularly struck with your remarks on Geog. Distrib. in Celebes. It is impossible that any thing c^{d} be better put, & w^{d} give a cold shudder to the immutable naturalists.[3]

And now I am going to ask a question which you will not like. How does y^{r} Journal get on?[4] It will be a shame if you do not popuralize your researches. my health is so far improved that I am able to work one or 2 hours a day—

Believe me dear Wallace | yours very sincerely | Ch. Darwin

1 A. R. Wallace, 'On the pigeons of the Malay Archipelago', *Ibis* n.s. 1 (1865): 365–400.

2 A. R. Wallace, 'On the phenomena of variation and geographical distribution as illustrated by the *Papilionidæ* of the Malayan region', *Transactions of the Linnean Society of London* 25 (1865–6): 1–71.

3 Compared with other islands of the archipelago, Celebes had many more species of Papilionidae that were unique to it; it also had 3 unique species of mammals, 5 of birds and 190 of hymenoptera (ants, wasps, bees, etc.). Wallace observed that members of the Papilionidae on Celebes exhibited certain common characteristics that distinguished them from their counterparts on other islands. He used this information to argue for the mutability of species and against the notion that all species had been created exactly as and where they were found.

4 It was not until 1867 that Wallace began in earnest to write *The Malay Archipelego*, his most popular book; it was published in 1869.

From A. R. Wallace 4 February 1866

9, S^{t}. Mark's Crescent, | *Regent's Park,* N.W.
Feb. 4th. *1866*

My dear Darwin

I am very glad to hear you are a little better, & hope we shall soon have the pleasure of seeing your volume on "Variation under domestication".

I do not see the difficulty you seem to feel about 2 or more female forms of one species. The *most common* or *typical* female form must have certain characters or qualities which are sufficiently advantageous to it, to enable it to maintain its existence;—in general such as vary much from it, die out. But occasionally a variation may occur which has special advantageous characters of its own, (such as mimicking a protected species) & then this variation will maintain itself by selection. In no less than 3 of my *polymorphic* species of Papilio, one of the female forms mimics the *Polydorus* group, which like the *Æneas* group in America seem to have some special protection. In two or three other cases one of the female forms is confined to a restricted locality to the conditions of which it is probably specially adapted. In other cases one of the female forms resembles *the male*, and perhaps receives a protection from the abundance of the males, in the crowd of which it is passed over.

I think these considerations render the production of two or three forms of female very conceivable. The physiological difficulty is to me greater, of how each of two forms of female, produces offspring like the other female as well as like itself, but no intermediates.

If you "*know varieties that will not blend or intermix, but produce offspring quite like either parent*",—is not that the very physiological test of a species which is wanting for the *complete proof* of the "origin of species:"?

I have by no means given up the idea of writing my travels, but I think I shall be able to do it better for the delay, as I can introduce chapters giving popular sketches of the subjects treated of in my various papers.

I hope, if things go as I wish this summer, to begin work at it next winter. But I feel myself incorrigibly lazy, & have no such system of collecting & arranging facts or of making the most of my materials, as you, & many of our hard working naturalists possess in perfection.

With best wishes | Believe me Dear Darwin | Yours most sincerely | Alfred R. Wallace

To A. R. Wallace [6 February 1866]

Down Bromley SE
Tuesday

My dear Wallace

After I had despatched my last note, the simple explanation which you give had occurred to me, & seems satisfactory.

I do not think you understand what I mean by the non-blending of certain varieties. It does not refer to fertility; an instance will explain; I crossed the Painted Lady & Purple sweet-peas, which are very differently coloured vars, & got, even out of the same pod, both varieties perfect but none intermediate. Something of this kind I sh[d] think must occur at first with your butterflies & the 3 forms of Lythrum; tho' these cases are in appearance so wonderful, I do not know that they are really more so than every female in the world producing *distinct* male & female offspring.

I am heartily glad that you mean to go on preparing your journal.

Believe me yours | very sincerely | Ch. Darwin

[In February 1866, Lyell, J. D. Hooker, Charles James Fox Bunbury, and CD began to circulate letters discussing recent claims by Louis Agassiz, arising out of his research expedition to Brazil, that the entire earth had been frozen during the Ice Age, destroying all life; this was a direct challenge to CD's theory of descent.]

To Charles Lyell 15 February [1866]

Down
Thursday | Feb 15th

My dear Lyell

Many thanks for Hooker's letter. It is a real pleasure to me to read his letters, they are alway written with such spirit. I quite agree that Agassiz could never mistake weathered-blocks & glacial action; though the mistake has, I know, been made in 2 or 3 quarters of the world. I have often fought with Hooker about the Physicists putting their veto on the world having been cooler; it seems to me as irrational, as if, when Geologists first brought forward some evidence of elevation & subsidence, a former Hooker had declared that this c[d] not possibly be admitted until Geologists c[d] explain what made the earth rise & fall. It seems that I erred greatly about some of the plants on the Organ M[ts]., but I am very glad to hear about Fuchsia &c. I cannot make out what Hooker does believe, he seems to admit the former cooler climate, & almost in the same breath, to spurn the idea. To retort Hooker's words "It is inexplicable to me" how he can compare the transport of seeds from the Andes to the Organs M[ts]. with that from a continent to an island: not to mention the much greater distance, there are no currents of water from one to the other, & what on earth sh[d] make a bird fly that distance without resting many times. I do not at all suppose that nearly all tropical forms were exterminated during the cool period, but in somewhat depopulated areas, into which there c[d] be no migration, probably many closely allied species will have been formed since this period. Hooker's paper in Nat. Hist. Rev. is *well worth* studying;[1] but I cannot remember that he gives good grounds for his conviction that certain orders of plants c[d] not withstand a rather cooler climate, even if it came on most gradually. We have only just learnt under how cool a temperature several tropical Orchids can flourish. I clearly saw Hookers difficulty about the preservation of tropical forms during the cool period, & tried my best to retain one spot after another as a hot-house for their preservation; but it w[d] not hold good, & it was a mere piece of truckling on my part when I suggested that longitudinal belts of the world were cooled one after the other. I shall very much like to see Agassiz' letter whenever you receive one.

I have written a long letter; but a squabble with or about Hooker always does me a world of good, & we have been at it many a long year. I cannot quite understand whether he attacks me as a Wriggler or a Hammerer but I am very sure that a deal of wriggling has to be done.

With many thanks | yours affectionately | Charles Darwin

1 CD apparently refers to an anonymous article, 'New colonial floras', *Natural History Review* n.s. 5 (1865): 46–63.

To John Murray 22 February [1866]

Down Bromley | Kent.
Feb 22—

My dear Sir

I am much pleased but even more grieved about the Origin;[1] for after ten months intermission I am now able to work nearly two hours daily at my next book;[2] but this will be now stopped by the Origin. Natural Hist. progresses so quickly that I must make a good many corrections. It will save me a good deal of labour if you will make a special request to Mess[rs] Clowes that the sheets are corrected with extra care, & only those sheets sent to me which contain corrections of more than a word or two. The former editions were corrected with *admirable* care. I will begin to work in a few days & as soon as a few sheets are ready shall I send them off to Mess[rs] Clowes?

I must cut up my present single copy, so must request you to send (to "care of G. Snow Nag's Head Borough")[3] a new bound copy as I must have one by me.

With respect to payment, will it suit you when half the copies are sold?

I hope you will let me have a few presentation copies as before.

I fear my Orchis book has been a poor affair. What state is it in?

I enclose a cheque for your account.

I was going to have written to you about woodcuts. Now Alas there is less hurry, but yet I may as well settle the affair. Ten blocks of pigeons & poultry are almost completed; but I require 32 or 33 of heads of animals—but chiefly of bones & skulls: there will often be 3 or 4 little bones in the same cut. Now I do not know that M[r] G. B. Sowerby has had any experience in drawing bones, but sh[d] rather prefer him as he is patient with me & I am familiar with his ways. What do you wish & advise? If M[r] Sowerby is employed, how is he, or indeed any one, to be restricted about price? Whoever draws for me will have to come down here to receive instructions & take away the specimens.

How long a time ought I to grant for these 32 woodcuts?

I am sorry to give you so much trouble with so many questions, & shall be grateful for answers & will give no more trouble.

I am very much interested in my present book on Domestic Animals &c; but cannot form the most remote idea whether the public will care for it. If it had not been for the Origin I think I sh[d] certainly have gone to press with it early this autumn.

I am much obliged for your kind enquiries about my health, & remain my dear Sir | yours very sincerely | Charles Darwin

P.S. I find my copy of Origin is so bescribbled, that I must correct on clean sheets— if you have unbound copy, so much the better.— *Please send by* **Post**.—

[1] CD refers to Murray's plans to print a fourth edition of *Origin*.

[2] CD had been engaged intermittently on *Variation* since January 1860.

[3] George Snow operated a carrier service between London and Down, calling at the Nag's Head inn, Borough High Street, London.

To J. D. Hooker 4 April [1866]

Down—
April 4th.

My dear Hooker

We have had G. Henslow here for two days & are very much pleased with him: there is something very engaging about him.—

Many thanks about the Bonatea & the Water-lilies & about the Cucumber case. Ask Mr Smith[1] whether by any odd chance he has ever seen a bud with blended character arising from junction of stock & graft.—

I will not forget about orchids; but it is not likely we shall have any to send you.— It was really **very** good in you to write about Pangenesis; for all such remarks lead one to see what points to bring out clearly.— I think you do not understand my notions on Pangenesis

Firstly.— I do not suppose that each cell can reproduce the whole species. The essence of my notion is that each cell, by throwing off an atom or gemmule (which grows or increases under proper conditions) reproduces the parent-cell & nothing more; but I believe that the gemmules of all the cells congregate at certain points & form ovules & buds & pollen-grains. I daresay they may congregate within a preexisting cell, passing through its walls like contents of pollen-tubes into embryonic sack; & it was partly on this account that I wished to learn about first appearance of buds.— When you speak of "a single detached cell of Begonia becoming a perfect plant"; I presume you do *not* mean that each cell, when separated by the knife, will grow; but that a fragment of a leaf will produce buds at apparently every & any point; if you mean more, I shd. be specially grateful for information.—

Secondly.— I do *not* suppose that gemmules are preserved in each species of all its preexisting states up to the "irrepressible monad"; but am forced to admit that wonderfully many are thus preserved & are capable of development, judging from reversion; but reversion does not go to such astounding lengths as you put it.

Thirdly. I do not suppose that a cell contains gemmules of any future state; but only that when a cell is modified by the action of the surrounding cells or of the external conditions, that the so modified cell throws off similarly modified atoms of its contents or gemmules which reproduce the modified cell.—

… I am not surprised that you think Pangenesis is only a statement of the concrete; so now it almost appears to me; yet I declare it has been nothing less than revelation to me as clearing away mist & connecting various classes of facts. The key-stone of the view is that the reproductive organs do not form the reproductive male & female elements,—only collect them (i.e. the gemmules of each separate cell) by some mysterious power in due proportions & fit them for mutual action & separate existence.—

If any remarks or sneers on this subject occur to you, for the love of Heaven, make a memorandum that I may sometime hear them.—

Ever yours affect. | C. Darwin

[1] Either John Smith (1821–88), the curator of the Royal Botanic Gardens, Kew, or John Smith (1798–1888), his predecessor.

To B. D. Walsh [19] April [1866]

Down. | Bromley. | Kent. S.E.
April 20th[1]

My dear Sir …

I see that you have been attacking Mr Scudder; & you will do the subject of the change of species wonderfully good service; for everyone in the U. States must now be aware that if he argues foolishly or misquotes, you will be down on him like a clap of thunder.[2] I have followed Sir C. Lyell's advice, (who is a very wise man) & always avoided controversy; but Lyell's arguments (except as far as loss of time is concerned) do not apply to any third party, who has energy & courage & wit enough to enter the arena.—

My health is considerably improved so that I work 2–3 hours daily; but all my new work has been stopped since the 1st of March, by correcting & adding to a new Edit. of the Origin. But I have found that I cd. not do nearly justice to the subject. I have referred to your work, but have not used it to *one quarter* of the extent, which I shd. have liked to have done. I will send you a copy when it is published in the course of the summer; for it is somewhat improved since the American Edition, which was so unfortunately stereotyped.— If you can remember look in Histor: Sketch at my account of Owen's views: it is rich & shows what a muddle those who "utter sonorous commonplaces about carrying out the Plan of Creation &c" fall into.—

My dear Sir | Yours most sincerely | Ch. Darwin …

My second son is now at your old College of Trinity, & has just gained a Scholarship, being the second man of his year, which pleases me much.—

[1] CD evidently misdated the letter: the cover is postmarked 19 April 1866.

[2] In a postscript to his paper 'On phytophagic varieties and phytophagic species', *Proceedings of the Entomological Society of Philadelphia* 3 (1864): 403–30, 5 (1865): 194–216, Walsh criticised a recent paper by Samuel Hubbard Scudder, in which a passage from *Origin* was misquoted.

From A. R. Wallace 2 July 1866

Hurstpierpoint, Sussex
July 2nd. 1866.

My dear Darwin

I have been so repeatedly struck by the utter inability of numbers of intelligent persons to see clearly or at all, the self acting & necessary effects of *Nat Selection*, that I am led to conclude that the term itself & your mode of illustrating it, however clear & beautiful to many of us are yet not the best adapted to impress it on the general *naturalist public*. The two last cases of this misunderstanding are, 1st. The article on "*Darwin & his teachings*" in the last "Quarterly Journal of Science", which, though very well written & on the whole appreciative, yet concludes with a charge of something like blindness, in your not seeing that "*Natural Selection*" requires the constant watching of an intelligent "chooser" like man's selection to which you so often compare it;—and 2nd., in Janet's recent work on the "*Materialism of the present day*",[1] reviewed in last Saturday's "Reader", by an extract from which I see that he considers your weak point to be, that you do not see that "*thought & direction are essential to the action of 'Nat. Selection'.*" The same objection has been made a score of times by your chief opponents, & I have heard it as often stated myself in conversation.

Now I think this arises almost entirely from your choice of the term "*Nat. Selection*" & so constantly comparing it in its effects, to *Man's selection*, and also to your so frequently personifying *Nature* as "*selecting*" as "*preferring*" as "*seeking only the good of the species*" &c.

&c. To the few, this is as clear as daylight, & beautifully suggestive, but to many it is evidently a stumbling block. I wish therefore to suggest to you the possibility of entirely avoiding this source of misconception in your great work, (if not now too late) & also in any future editions of the "Origin", and I think it may be done without difficulty & very effectually by adopting Spencer's term (which he generally uses in preference to Nat. Selection) viz. "*Survival of the fittest.*"[2]

This term is the plain expression of the *facts*,—*Nat. selection* is a metaphorical expression of it—and to a certain degree *indirect* & *incorrect*, since, even personifying Nature, she does not so much *select* special variations, as *exterminate* the most unfavourable ones.

Combined with the enormous multiplying powers of all organisms, & the "struggle for existence" leading to the constant destruction of by far the largest proportion,—facts which no one of your opponents, as far as I am aware, has denied or misunderstood,—"*the survival of the fittest*" rather than of those who were less fit, could not possibly be denied or misunderstood. Neither would it be possible to say, that to ensure the "*survival of the fittest*" any *intelligent chooser* was necessary,—whereas when you say *natural selection* acts so as to choose those that are fittest it *is* misunderstood & apparently always will be. Referring to your book I find such expressions as "Man selects only for his own good; Nature only for that of the being which she tends". This it seems will always be misunderstood; but if you had said "Man selects only for his own good; Nature, by the inevitable "survival of the fittest", only for that of the being she tends",—it would have been less liable to be so.

I find you use the term "Natural Selection" in two senses, 1st for the simple preservation of favourable & rejection of unfavourable variations, in which case it is equivalent to "*survival of the fittest*",—or 2nd. for the *effect* or *change*, produced by this preservation, as when you say, "To sum up the circumstances favourable or unfavourable to *natural selection*", and again "Isolation, also, is an important element in the process of *natural selection*",—here it is not merely "*survival of the fittest*" but, *change produced by survival of the fittest*, that is meant— On looking over your fourth Chap. I find that these alterations of terms can be in most cases easily made, while in some cases the addition of "*or survival of the fittest*", after "*natural selection*" would be best; and in others, less likely to be misunderstood, the original term may stand alone.

I could not venture to propose to any other person so great an alteration of terms, but you I am sure will give it an impartial consideration, and if you really think the change will produce a better understanding of your work, will not hesitate to adopt it.

It is evidently also necessary not to *personify* "nature" too much,—though I am very apt to do it myself,—since people will not understand that all such phrases are metaphors.

Natural selection, is, when understood, so necessary & self evident a principle, that it is a pity it should be in any way obscured; & it therefore occurs to me, that the free use of "*survival of the fittest*",—which is a compact & accurate definition of it,—would tend much to its being more widely accepted and prevent its being so much misrepresented & misunderstood.

There is another objection made by Janet which is also a very common one. It is that the chances are almost infinite again the particular kind of variation required being coincident with each change of external conditions, to enable an animal to become modified by Nat. Selection in harmony with such changed conditions; especially when we consider, that, to have produced the almost infinite modifications of organic beings this coincidence must have taken place an almost infinite number of times.

Now it seems to me that you have yourself led to this objection being made, by so often stating the case *too strongly* against yourself. For Example, at the Commencement of Chap. IV. you ask, if it is "*improbable that useful variations should sometimes occur in the course of thousands of generations*";—and a little further on you say, "*unless profitable variati⟨ons⟩ do occur natural selection can do nothing.*" Now such expressions h⟨ave⟩ given your opponents the advantage of assuming that *favourable variations* are *rare accidents*, or may even for long periods never occur at all, & thus Janet's argument would appear to many to have great force. I think it would be better to do away with all such qualifying expressions, and constantly maintain (what I certainly believe to be the fact) that *variations* of *every kind* are *always occurring* in *every part* of *every species*,—& therefore that favourable variations are *always ready* when wanted. You have I am sure abundant materials to prove this, and it is, I believe, the grand fact that renders modification & adaptation to conditions almost always possible. I would put the burthen of proof on my opponents, to show, that any one *organ structure* or *faculty* does *not vary*, even during one generation among all the individuals of a species,—and also to show *any mode or way* in which any such *organ* &c. does not vary. I would ask them to give any reason for supposing that *any organ* &c. is ever *absolutely identical* at any *one time* in *all the individuals* of a species,—& if not then it is always varying, and there are always materials which, from the simple fact, that "*the fittest survive*", will tend to the modification of the race into harmony with changed conditions.

I hope these remarks may be intelligible to you, & that you will be as kind as to let me know what you think of them.

I have not heard for some time how you are getting on.

I hope you are still improving in health, & that you will be able now to get on with your great work for which so many thousands are looking with interest.

With best wishes | Believe me My dear Darwin | Yours very faithfully | Alfred R. Wallace— . . .

[1] Paul Janet, *The materialism of the present day* (London, 1866).

[2] Herbert Spencer first used the expression 'survival of the fittest' in *Principles of biology* (London, 1864–7), 1: 444–5: 'This survival of the fittest, which I have here sought to express in mechanical terms, is that which Mr. Darwin has called "natural selection, or the preservation of favoured races in the struggle for life".'

To A. R. Wallace 5 July [1866]

Down. | Bromley. | Kent. S.E.

July 5th.

My dear Wallace

I have been much interested by your letter which is as clear as daylight. I fully agree with all that you say on the advantages of H. Spencer's excellent expression of "the survival of the fittest. This however had not occurred to me till reading your letter. It is, however, a great objection to this term that it cannot be used as a substantive governing a verb; & that this is a real objection I infer from H. Spencer continually using the words natural selection.

I formerly thought, probably in an exaggerated degree, that it was a great advantage to bring into connection natural & artificial selection; this indeed led me to use a term in common, and I still think it some advantage. I wish I had received your letter two months ago for I would have worked in "the survival etc" often in the new edition of the Origin which is now almost printed off &

of which I will of course send you a copy. I will use the term in my next book on Domestic Animals etc from which, by the way, I plainly see, that you expect *much* too much.[1] The term Natural selection has now been so largely used abroad & at home that I doubt whether it could be given up, & with all its faults I should be sorry to see the attempt made. Whether it will be rejected must now depend "on the survival of the fittest". As in time the term must grow intelligible, the objections to its use will grow weaker & weaker. I doubt whether the use of any term would have made the subject intelligible to some minds, clear as it is to others; for do we not see even to the present day Malthus on Population absurdly misunderstood. This reflexion about Malthus has often comforted me when I have been vexed at the misstatement of my views. As for M. Janet he is a metaphyscian & such gentlemen are so acute that I think they often misunderstand common folk. Your criticism on the double sense in which I have used Natural Selection is new to me and unanswerable; but my blunder has done no harm, for I do not believe that anyone excepting you has ever observed it. Again I agree that I have said too much about "favourable variations;" but I am inclined to think that you put the opposite side too strongly: if every part of every being varied, I do not think we should see the same end or object gained by such wonderfully diversified means.

I hope you are enjoying "the country & are in good health", and are working hard at your Malay Arch. book, for I will always put this wish in every [note] I write to you, like some good people always put in a text.

My health keeps much the same or rather improves & I am able to work some hours daily.

With many thanks for your interesting letter, believe me, | my dear Wallace, yours sincerely | Ch. Darwin

P.S. I suppose you have read the last number of H. Spencer; I have been struck with astonishment at the prodigality of Original thought in it; but how unfortunate it is that it seems scarcely ever possible to discriminate between the direct effect of external influences & "the survival of the fittest".—

[1] CD used the expression 'survival of the fittest' six times in *Variation* (see *Variation* 1: 6, 2: 89, 192, 224, 413, 432); however, CD also defended his use of 'natural selection' in *Variation* 1: 6: 'The term "natural selection" is in some respects a

bad one, as it seems to imply conscious choice; but this will be disregarded after a little familiarity.'

To J. D. Hooker 3 and 4 August [1866]

Down.
Aug. 3rd.

My dear Hooker

Manny thanks for Acropera & about the book. I will take your letter seriatim. There is good evidence that S.E. England was dry land during glacial period. I forget what Austin[1] says but mammals prove, I think, that England has been united to the Continent since the glacial period. I don't see your difficulty about what I say on the breaking of an isthmus: if Panama was broken thro', wd. not the fauna of the Pacific flow into the W. Indies, or vice versâ, & destroy a multitude of creatures. Of course I'm no judge, but I thought De Candolle had made out his case about small areas of trees— You will find at P. 112 3rd. edit. Origin a too concise allusion to the Madeira flora being a remnant of the tertiary European flora.

I shall feel deeply interested by reading your botanical difficuties against occasional immigration. The facts you give about certain plants such as the Heaths are certainly very curious. I thought the Azores flora was more boreal: but what can you mean by saying that the Azores are nearer to Britain & Newfoundland than to Madeira? on the Globe they are nearly twice as far off. With respect to sea-currents, I formerly made enquiries at Madeira but cannot now give you the results but I remember that the facts were different from what is generally stated; I think that a ship wrecked on the Canary Islands was thrown up on the coast of Madeira. You speak as if only land shells differed in Madeira & Po. Santo: does my memory deceive me that there is a host of representative insects?

When you exorcise at Nottingham occasional means of transport, be honest, & admit how little is known on the subject.[2] Remember how recently you & others thought that Salt-water would soon kill seeds. Reflect that there is not a coral-islet in the ocean which is not pretty well clothed with plants: & the fewness of the species can hardly with justice be attributed to the arrival of few seeds, for coral-Islets close to other land support only the same limited vegetation. Remember that no one knew that seeds wd remain for many hours in the *crops* of birds & retain their vitality; that fish eat seeds & that when the fish are devoured by birds the seeds can germinate &c &c— Remember that every year many birds are blown to Madeira & to the Bermudas. Remember that dust is blown 1000 miles over the Atlantic. Now bearing all this in mind, wd it not be a prodigy if an *unstocked* Island did not in the course of ages receive colonists from coasts, whence the currents flow, trees are drifted, & birds are driven by gales. The objections to islands being thus stocked are, as far as I understand, that certain species & genera have been more freely introduced & others less freely than might have been expected. But then the sea kills some sorts of seeds, others are killed by the digestion of birds & some wd be more liable than others to adhere to birds feet; but we know so very little on these points that it seems to me that we cannot at all tell what forms wd probably be introduced & what wd not.

I do not for a moment pretend that these means of introduction can be proved to have acted; but they seem to me sufficient, with no valid or heavy objections, whilst there are, as it seems to me, the heaviest objections, on geological & on geographical-Distribution grounds, (p. 387, 388, Origin) to Forbes' enormous continental extensions. But I fear that I shall & have bored you.—

Yours ever affect— | C. Darwin.

P.S. Murray will not bring out, & be hanged to him, the new Edit. of Origin, though all printed off, till November.— I have persuaded him to send Lyell a copy; I do not suppose you wd. care to have your copy at once; if you did, I would ask Murray, but for some reason, he does not seem much to like sending out the copies.—

Dont answer unless you like, for you must be very busy.—

P.S. Here is a bad job, the Acropera has not arrived.— I hope it was not sent off, as soon as you thought. I sent this evening (Friday) but no parcel at Station. If not there tomorrow it must be lost. It is a bad job for me & for you, if the plant is valuable.—

P.S. 2d. As you were asking about Books on "Origin"; a very good Zoologist Claus has just published one, with my name on title-page—the subject being an investigation of the amount of individual variability in the Copepodous Crustaceans & he shows it is wonderfully great in many organs & that some *co existing* vars, are apparently passing into distinct species.—

☞ **Acropera** all safe Saturday morning

1 Robert Alfred Cloyne Godwin-Austen.

2 Hooker was to give a lecture on the geographical distribution of plants at the meeting of the British Association for the Advancement of Science in Nottingham (see 'On insular floras: a lecture', *Journal of Botany* 5 (1867): 23–31).

To J. D. Hooker 8 August [1866]

Down. | Bromley. | Kent. S.E.
Aug 8

My dear Hooker

It w^{d} be a very great pleasure to me if I c^{d} think that my letters were of the least use to you. I must have expressed myself badly for you to suppose that I look at islands being stocked by occasional transport as a well established hypothesis: we both give up creation & therefore have to account for the inhabitants of islands either by continental extensions or by occasional transport; now all that I maintain is that of these two alternatives, one of which must be admitted notwithstanding very many difficulties, that occasional transport is by far the most *probable.*

I go thus far further that I maintain, knowing what we do, that it w^{d}. be inexplicable if *unstocked* isl^{ds} were not stocked to certain extent at least, by these occasional means.—

European birds are occasionally driven to America but far more rarely than in the reverse direction: they arrive viâ Greenland (Baird):[1] yet a European lark has been caught in Bermuda. By the way you might like to hear that European birds regularly migrate, viâ the Northern Islands, to Greenland. . . .

I do not think it a mystery that birds have not been modified in Madeira. Pray look at p. 422 of Origin. You w^{d} not think it a mystery if you had seen the long lists which I have (somewhere) of the birds annually blown, even in flocks, to Madeira. The crossed stock would be the more vigorous.—

Remember if you do not come here before Nottingham, if you do not come afterwards I shall think myself diabolically ill-used.

yours affectionately | Ch. Darwin

P.S. Ought you not to measure from the Azores, not to Newfoundland, but to the more Southern & temperate States?

[1] S. F. Baird, 'The distribution and migrations of North American birds', *American Journal of Science and Arts* 41 (1866): 344–5.

To William Darwin Fox 24 August [1866]

Down. | Bromley. | Kent. S.E.
Aug 24th.

My dear Fox

It is always a pleasure to me to hear from you & about you, patriarch as you are with your countless children, & grand children—[1] I feel a mere dwarf by your side. My own children are all well & two of my boys are touring in Norway.[2] Poor Susan is in a terrible suffering state & I fear there is no hope for her except the one & last hope for all.[3]

We expect Caroline [Wedgwood] here with her three girls on Monday & I will give her your kind enquiries. As for myself my health is very decidedly better though I am not very strong. I attribute my improvement partly to Bence Jones' diet & partly, wonderful to relate, to my riding every day which I enjoy much. I don't believe in your theory of moderate mental work doing me any harm—any how I can't be idle. I am making rapid progress with my book on domesticated animals which I fear will be a big one & has been laborious from the number of references. I hope to begin printing towards the close of this close of this year & when it is completed I will of course send you a copy, as indeed I am bound to do as I owe much information to you. I should have begun printing before this had I not lost nearly three months by the troublesome labour of largely correcting a new edit. of the Origin. I think I have heard of hybrids between the Sphinx-moths which you mention; I shall be surprised if the hybrids are fertile

Believe me, my dear old Friend | yours affectly | Charles Darwin

[1] Fox had sixteen children and five grandchildren.

[2] George Howard Darwin and Francis Darwin

[3] CD's sister, Susan Elizabeth Darwin, died on 3 October 1866.

[One of CD's most assiduous scientific correspondents was Fritz Müller, a German naturalist living in Brazil. In 1864, he introduced himself by sending CD a copy of his book, *Für Darwin* (Leipzig, 1864), a study of the Crustacea with reference to CD's theory of transmutation; CD later financed an English translation. CD in due course sent Müller a copy of his 1865 paper 'Climbing plants', after which Müller sent CD a wealth of observations on climbing plants near his own home. CD collated Müller's letters and had them published in the *Journal of the Linnean Society* for 1866. Also in 1866, CD asked Müller about orchids in his vicinity. He called Müller 'the prince of observers'.]

To Fritz Müller 25 September [1866]

Down Bromley Kent
Sep 25th

My dear Sir

I have just recd your letter of Aug. 2nd & am as usual astonished at the number of interesting points which you observe. It is quite curious how by coincidence you have been observing the same subjects that have lately interested me.

Your case of the Notylia is quite new to me; but it seems analogous with that of Acropera, about the sexes of which I blundered greatly in my book. I have got an Acropera now in flower & have no doubt that some insect with a tuft of hairs on its tail removes by the tuft the pollinia, & inserts the little viscid cap & the long pedicel into the narrow stigmatic cavity, & leaves it there with the pollen-masses in close contact with, but not inserted into, the stigmatic cavity. I find I can thus fertilize the flowers; & so I can with Stanhopea, & I suspect that this is the case with your Notylia. But I have lately had an orchid in flower, viz. Acineta, which I could not any how fertilize. Dr Hildebrand lately wrote a paper shewing that with some orchids the ovules are not mature & are not fertilized until months after the pollen-tubes have penetrated the column; & you have independently observed the same fact, which I never suspected in the case of Acropera. The column of such orchids must act almost like the Spermatheca of insects. Your Orchis with 2 leaf-like stigmas is new to me; but I feel guilty at your wasting your valuable time in making such beautiful drawings for my amusement.

Your observations on those plants being sterile which grow separately or flower earlier than others are very interesting to me.

They w[d] be worth experimenting on with other individuals: I shall give in my next book several cases of *individual* plants being sterile with their own pollen. I have actually got on my list Escholtzia for fertilizing with its own pollen, though I did not suspect it w[d] prove sterile, & I will try next summer. My object is to compare the rate of growth of plants raised from seed fertilized by pollen from the same flower & by pollen from a distinct plant & I think from what I have seen I shall arrive at interesting results. D[r] Hildebrand has lately described a curious case of Corydalis cava, which is quite sterile with its own pollen, but fertile with pollen of *any* other individual plant of the species. What I meant in my paper on Linum about plants being dimorphic in function alone was that they sh[d] be divided into two equal bodies functionally but not structurally different. I have been much interested by what you say on seeds which adhere to the valves being rendered conspicuous: you will see in the new Edit. of the origin why I have alluded to the *beauty* & bright colours of fruit; after writing this, it troubled me that I remembered to have seen brilliantly coloured seed, & your view occurred to me. There is a species of Peony in which the inside of the pod is crimson & the seeds dark purple. I had asked a friend to send me some of these seeds, to see if they were covered with any thing which c[d] prove attractive to birds. I rec[d] some seeds the day after receiving your letter; & I must own that the fleshy covering is so thin that I can hardly believe it w[d] lead birds to devour them; & so it was in an analogous case with Passiflora gracilis. How is this in the cases mentioned by you? The whole case seems to me rather a striking one.

I wish I had heard of Mikania being a leaf-climber before your paper was printed;[1] for we thus get a good gradation from M. scandens to Mutisia with its little modified leaf-like tendrils. I am glad to hear that you can confirm (but render still more wonderful) Haeckel's most interesting case of Liriope: Huxley told me that he thought that the case w[d] somehow be explained away.

As for Agassiz & his glaciers in the valley of the Amazons, it seems to me sheer madness, as it does likewise to Lyell; the evidence being wholly insufficient. Prof. Asa Gray tells me that A. started with the determination to prove that the whole world had been covered with ice in order to annihilate all Darwinian views.

I hope I have not troubled you with this long letter & believe me yours very sincerely
Charles Darwin

[1] 'Notes on some of the climbing-plants near Desterro, in south Brazil. By Herr Fritz Müller, in a letter to C. Darwin', *Journal of the Linnean Society (Botany)* 9 (1866): 344–9.

To Charles Lyell 9 October [1866]

Down.
Oct 9th

My dear Lyell

One line to say that I have received your note & the Proofs safely, & will read them with greatest pleasure;[1] but I am *certain* I shall not be able to send any criticism on the Astronomical Chapter, as I am as ignorant as a pig on this head.

I shall require some days to read what has been sent. I have just read Chapter IX & like it *extremely*: it all seems to me very clear, cautious & sagacious.[2] You do not allude to one very striking point enough or at all, viz the classes having been formerly less differentiated than they now are; & this specialisation of classes must, we may conclude, fit them for different general habits of life, as well as the specialisation of particular organs.— …

You will think me impudent, but the discussion at end of Ch IX on man, who thinks so much of his fine self, seems to me too long or rather superfluous & too orthodox, except for the beneficed clergy.—

Ever yours | C. Darwin

[1] CD refers to proofs of the first volume of the tenth edition of Lyell's *Principles of geology* (London, 1867–8).

[2] Chapter 9 of the *Principles of geology*, which dealt with the progressive development of organic life, had been entirely re-written.

[Mary Everest Boole was employed as a librarian in Queen's College, Harley Street, London, the first women's college in England. Although she had no formal teaching duties, she gave Sunday evening talks in which she discussed the relationship of different forms of knowledge. She was especially interested in the psychology of learning and her ideas on child psychology and learning were later taken up by educators in America.]

From Mary Everest Boole 13 December 1866

Private

Dear Sir

Will you excuse my venturing to ask you a question to which no one's answer but your own would be quite satisfactory to me.

Do you consider the holding of your Theory of Natural Selection, in its fullest & most unreserved sense, to be inconsistent,—I do not say with any particular scheme of Theological doctrine,—but with the following belief, viz:

That knowledge is given to man by the direct Inspiration of the Spirit of God.

That God is a personal and Infinitely good Being.

That the effect of the action of the Spirit of God on the brain of man is *especially* a moral effect.

And that each individual man has, within certain limits, a power of choice as to how far he will yield to his hereditary animal impulses, and how far he will rather follow the guidance of the Spirit Who is educating him into a power of resisting those impulses in obedience to moral motives.

The reason why I ask you is this. My own impression has always been,—not only that your theory was quite *compatible* with the faith to which I have just tried to give expression,—but that your books afforded me a clue which would guide me in applying that faith to the solution of certain complicated psychological problems which it was of practical importance to me, as a mother, to solve. I felt that you had supplied one of the missing links,—not to say *the* missing link,—between the facts of Science & the promises of religion. Every year's experience tends to deepen in me that impression.

But I have lately read remarks, on the probable bearing of your theory on religious & moral questions, which have perplexed & pained me sorely. I know that the persons who make such remarks must be cleverer & wiser than myself. I cannot feel sure that they are mistaken unless you will tell me so. And I think,—I cannot know for certain, but I *think*,—that, if I were an author, I would rather that the humblest student of my works should apply to me directly in a difficulty than that she should puzzle too long over adverse & probably mistaken or thoughtless criticisms.

At the same time I feel that you have a perfect right to refuse to answer such questions as I have asked you. Science must take her path & Theology hers, and they will meet when & where & how God pleases, & you are in no sense responsible for it, if the meeting-point should be still very far off. If I receive no answer to this letter, I shall infer nothing from your silence except that you felt I had no right to make such inquiries of a stranger.

I remain | Dear Sir | Yours truly | Mary Boole

To M. E. Boole 14 December 1866

Down. Bromley. Kent.
Decr. 14. 1866.

Dear Madam.

It would have gratified me much if I could have sent satisfactory answers to y^{r}. questions, or indeed answers of any kind. But I cannot see how the belief that all organic beings including man have been genetically derived from some simple being, instead of having been separately created bears on your difficulties.— These as it seems to me, can be answered only by widely different evidence from Science, or by the so called "inner consciousness". My opinion is not worth more than that of any other man who has thought on such subjects, & it would be folly in me to give it; I may however remark that it has always appeared to me more satisfactory to look at the immense amount of pain & suffering in this world, as the inevitable result of the natural sequence of events, i.e. general laws, rather than from the direct intervention of God though I am aware this is not logical with reference to an omniscient Deity— Your last question seems to resolve itself into the problem of Free Will & Necessity which has been found by most persons insoluble.

I sincerely wish that this note had not been as utterly valueless as it is; I would have sent full answers, though I have little time or strength to spare, had it been in my power.

I have the honor to remain dear Madam. | Yours very faithfully | Charles Darwin.

P.S. I am grieved that my views should incidentally have caused trouble to your mind but I thank you for your Judgment & honour you for it, that theology & science should each run its own course & that in the present case I am not responsible if their meeting point should still be far off.

From M. E. Boole 17 December [1866]

43 Harley S^{t}.
Decr. 17th.

Dear Sir,

Thank you sincerely for your kind letter. You have told me all I wanted to know from you. The criticisms to which I referred were such as seemed to take for granted that all such speculations as yours,—in fact, as it seemed to me, *all* independent untheological speculations on Creation as we find it,—must be incompatible with any belief in a moral government of the world. I have always taken the liberty of telling the people who brought such criticisms under my notice, that, in my opinion, the authors of them were simply talking about what they had never examined into. But still, when one is studying alone, & so ignorant too as I am, one gets frightened, & loses faith in one's own principles. And I thought, for my own satisfaction, I should like to have *your* assurance that moral & religious faith are things quite independent of theories about the *process* of Creation. You have given me that assurance and again I thank you.

With sincere wishes for improvement in your health

I remain | dear Sir | Yours truly | Mary Boole

From Lydia Ernestine Becker 22 December 1866

10 Grove st | Ardwick. Manchester
Dec. 22. 1866.

My dear M^{r}. Darwin

Before proceeding to the object of my letter I must try to recall my name to your recollection. I scarcely dare flatter myself, that I can do this successfully, though the remembrance of your great kindness and courtesy to me will never fade from my mind and is a constant source of pride and pleasure.

In the summer of 1863 I ventured to send you some flowers of *Lychnis diurna* which seemed to present some curious characteristics, and though they proved on examination not to possess the interest you at first thought they might have with respect to your own investigations you were good enough to write me several notes about them. You also did me the honour to send me a copy of a paper you had read to the Linnæan Society on two forms in the genus *Linum* and I had the greatest pleasure in immediately procuring a pot of seedlings of crimson flax—and watching the appearances you had recorded

At first the plants which bloomed had long styled flowers only, and at this point my correspondence with you came to a natural close, leaving you under the impression that there had been a failure in my observations. I have often since wished that I could have expressed to you the admiration and delight with which I perceived how, when the short styled flowers at last made their appearance, the capsules which till then, had withered away with the petals, seemed to start into life—how they grew and they swelled and rapidly became vigorous and healthy fruits— But I had no pretext for troubling you with any further communication and I feared I had already trespassed too much on your attention.

I have not been able to pursue my study of the Lychis flowers nor my endeavours to penetrate the mystery of their alteration in form, for since then we have ceased to reside in the country and now, surrounded by acres of bricks and mortar—and an atmosphere laden with coal smoke, I have no opportunity of watching living plants.

But living in a town has its advantages, among others it makes possible such societies as that indicated in the circular I have taken the liberty to enclose. A few ladies have joined together hoping for much pleasure and instruction from their little society, which is quite in its infancy and needs a helping hand. Am I altogether too presumptuous in seeking this help from you? Our petition is—would you be so very good as to send us a paper to be read at our first meeting Of course we are not so unreasonable as to desire that you should write anything specially for us, but I think it possible you may have by you a copy of some paper such as that on the *Linum* which you have communicated to the learned societies but which is unknown and inaccessible to us unless through your kindness. In your paper on the *Linum* you mention your experiments on *Primula* which greatly excite my interest and curiosity, for last spring as I was gathering primroses I was forcibly struck with the difference between the "pin eyed" flowers, and those in which the stigma was concealed beneath the anthers. I have known of this difference in the Polyanthus from childhood, but not until I read your paper was I aware of its interest or importance and now I have just enough information to excite and tantalise, but not to satisfy, a strong desire for more. If you will pardon the presumption of the request, I would beg that your goodness might prompt you to send something you may have on hand in the form of pamphet or paper which would help us to learn the meaning of these curious differences in the flowers, and as we may all hope during the coming spring for the pleasure of luxuriating on a primrose bank we should indeed be grateful for the kindness that had guided us to look more closely into the beautiful things we were enjoying.

I send this with much misgiving lest you may be displeased at the liberty I have taken if I have a hope of pardon it rests entirely on your goodness.[1]

Believe me always | yours much obliged | Lydia E. Becker.

[1] CD sent 'Climbing plants' and 'Three forms of *Lythrum salicaria*'. See letter from L. E. Becker, 6 February 1867.

To J. D. Hooker 24 December [1866]

Down
Dec 24

My dear Hooker.

I am going to amuse myself by scribbling a bit to you about your last long letter. But first you must congratulate me in your mind when you hear that I have sent M.S. (such an awful, confounded pile, two volumes I much fear) of "Domestic Animals & Cult. Plants" to Printers.[1] I am now writing concluding chapter, & shall perhaps insert, but am much perplexed on this head, a Chapt. on Man; just to say how I think my views bear on him.—

We have all the Boys at home & are very jolly, & William has come for 3 days. He has brought back the Introduction to Australian Flora,[2] after having read it over *three* times & liked it *extremely*. I mention this because it shows how interesting & valuable a book you might produce for general readers on Insular Floras. I feel, however, sure that you will grapple with this work now.— I see in Müllers letter that I assumed without any grounds that the Adenanthera was a native Brazilian plant: it is not worth enquiring in India about, though it is a perplexing case, for I can hardly admit your wriggle of the seeds being devoured by birds with weak gizzards: at least soaking for 10 hours in a little warm water got out hardly anything soluble from one of the seeds. Yet I must believe that they hang long on the tree & look so gaudy to attract birds.—

I read aloud your simile of H. Spencer to a thinking pump, & it was unanimously voted first-rate, & not a bit the worse for being unintelligible.

One word more about about the flora derived from supposed pleistocene Antarctic Land, requiring land intercommunication, this will depend much, as it seems to me, upon how far you finally settle whether Azores, Cape de Verdes, Tristan d'Acunha, Galapagos Juan Fernandez &c &c &c have all had land intercommunication. If you do not think this necessary might not New Zealand &c have been stocked during intervening glacial period by occasional means from Antarctic Land? As for lowlands of Borneo being tenanted by a moderate number of temperate forms during Glacial period, so far from appearing a "frightful assumption", that I am arrived at that pitch of bigotry that I look at it as *proved*!

I had another letter from Fritz Müller yesterday & in one day's collecting he found six genera of dimorphic plants! One is a Plumbago.— Now have you seed of any species; I see none are on sale in Carter's list.[3] I want a second favour; could you lend me for *short time* a recent number of Revue Horticole, with an account by Carrière of curious effect of grafting an Aria, given in last Gard. Chronicle.—[4]

Yours affect | C. Darwin

[1] *Variation.*

[2] J. D. Hooker, *On the flora of Australia, its origin, affinities, and distribution; being an introductory essay to the flora of Tasmania* (London, 1859).

[3] CD refers to a list of seeds from the firm Carter, Dunnett & Beale; the list had been published annually since 1837.

[4] E. A. Carrière, 'Transformation de l'Aria vestita par la greffe', *Revue Horticole* 37 (1866): 457–8.

1867

To John Murray 3 January [1867]

Down. | *Bromley.* | *Kent. S.E.*
Jan. 3[d]

My dear Sir

I cannot tell you how sorry I am to hear of the enormous size of my Book.[1] I fear it can never pay. But I cannot shorten it now; nor indeed, if I had foreseen its length, do I see which parts ought to have been omitted.

If you are afraid to publish it, say so at once I beg you, & I will consider your note as cancelled. If you think fit get anyone whose judgment you rely on, to look over some of the more legible chapters; viz the Introduction & on Dogs & on Plants; the latter chapters being in my opinion the dullest in the book. There is a Hypothetical & curious Chapter called Pangenesis which is legible, & about which I have no idea what the instructed public will think; but to my own mind it has been a considerable advance in knowledge— The list of Chapters, & the inspection of a few, here & there, w[d] give a good judge a fair idea of the whole Book. Pray do not publish blindly, as it would vex me all my life if I led you to heavy loss. I am extremely much vexed at the size; but I believe the work has some value, though of course I am no fair judge.—

You must settle all about type & size according to your own judgment; but I will only say that I think, *& hear on all sides* incessant complaint of the fashion which is growing of publishing intolerably heavy volumes:—

I have written my concluding Chapter; whether that on Man, shall appear, shall depend on size of book, on time & on my own strength.

My dear Sir | Yours very sincerely | Ch. Darwin

[1] CD refers to *Variation.*

From John Murray 28 January [1867]

Albemarle St
Jany 28

My Dear Sir

Pray put yourself at ease about the publication of your new book. I will publish it for you coute qui coute[1] provided you will be content that I pay you one half the profits of the edition instead of a sum down at first— This I ask because—no doubt there is considerably greater risque in this than in the publication of your former works.—

This work is not intended nor likely to become generally popular but I think after the sale of 6000 of your "Origins" I can *count* upon 500 purchasers of these new volumes—the "Pièces Justificatives" on wch that work is founded & I w^{d} propose to print an Edition of 750 copies—in the size type & page of Lyells Principles—like wch it will make 2 volumes 8vo.

I have heard from my literary friend—but have not yet got back the MS.S from him— He certainly finds it difficult of digestion but he is not a man of science so his opinion is not a fair test altogether— Still in the face of it, I venture to submit to you the above proposal.

I am My Dear Sir | Yours very faithfully | John Murray

I hope to return the MS. this week | JM

[1] *Coûte que coûte*: at any cost.

From L. E. Becker 6 February 1867

Manchester Ladies' Literary Society. | 10 Grove s^{t} | Ardwick
Feb. 6. 1867.

My dear Sir

I return you—with more thanks than I know how to express, the two papers which you were so good as to entrust to my care. . . .[1]

I have transcribed portions of them, and made large copies of the diagrams— . . .

The arrangements in *Lythrum* are indeed most marvellous. It sets one wondering whether different sized stamens in the same flower can ever be quite without meaning, and if there is any difference in the action of the pollen of the long and short stamens in didynamous and tetradynamous flowers. In the N. O.

Geraniaceae it seems as if there might be some transition going on—for in *Geranium* each alternate stamen is smaller, and in the allied genus *Erodium* the alternate stamens have become sterile. Can it be possible that this genus was once dimorphic, and one of the female forms having by any means become exterminated, the corresponding set of stamens have shed away? If one of the forms of Lythrum were to disappear—two sets of stamens would be made useless to the species, and it is conceivable that they might then gradually become abortive.

I obeyed your directions about the paper on Climbing Plants and the insight into their extraordinary and regular movements was a new revelation to all of us. I made large copies of the diagrams and dived into my herbarium for specimens of each class of climbers, bringing up enough to make a goodly show. Luckily a collection of ferns from the islands of the South Pacific recently presented to me contained a specimen of one named in your paper *Lygodium scandens*. Till I read it I had never dreamed of twiners in this class, as none of our British ferns have the habit, but as the "march of intellect" seems to be the order of the day, even in the vegetable world, there is no telling what they may accomplish in time!

Our society appears likely to prosper beyond my expectations the countenance you have afforded has been of wonderful service, and I do hope that by becoming useful to its members it may prove in some degree worthy of the generous encouragement you have given us.

The ladies who had the privilege of listening to the paper desire to express their thanks to you for it, which I hope you will be pleased to accept.

Believe me to be | yours gratefully | Lydia E. Becker.

[1] 'Climbing plants' and 'Three forms of *Lythrum salicaria*'.

To W. D. Fox 6 February [1867]

Down Bromley Kent
Feb 6th

My dear Fox

It is always a pleasure to me to hear from you, & old & very happy days are thus recalled. This is rather a joyful day to me, as I have just sent off the M.S for two huge volumes (I grieve the Book is so big) to Printers on Domestic Animals &c &c but my book will not appear, even if completed, before next November, as Murray has strong prejudice against publishing except during Spring & Autumn. I am utterly in darkness about merit of my present book; all that I know is that it has been a most laborious undertaking. Of course a copy will be sent to you.—

It is true indeed that Death has been busy with us, & it is astonishing to me that I sh^{d}. have survived my two poor dear sisters. The old House at Shrewsbury is on sale, but has as yet found no purchaser, & I daresay will not soon.— All the furniture was sold by Auction, having been bequeathed to the Parkers,[1] who had become like Susan's children.

Caroline & Erasmus are fairly well for them; but this is not saying much for them, especially for the latter, who does not often leave the House. I am so sorry to hear so poor an account of yourself; with your active habits being confined must be a terrible deprivation. You are quite right about riding; it does suit me admirably, & I am very much stronger; yet I never pass 12 hours without much energetic discomfort. But I am fairly well content, now that I am no longer quite idle.— Poor Bence Jones has been for months at death's door, & was quite given up; but has rallied in surprising manner from inflammation of Lungs & heart-disease. My wife is fairly well but suffers much from repeated headachs, & the rest of us are well.— I hope you will get all right with returning Spring.

My dear old friend | Believe me; | Yours affectionately | Ch. Darwin

[1] CD's eldest sister, Marianne, married Henry Parker (1788–1856); after the Parker parents died, their grown-up children had stayed at The Mount when they were in Shrewsbury.

To J. D. Hooker 8 February [1867]

Down Bromley Kent
Feb 8th

My dear Hooker

I am heartily glad that you have been offered the Presidentship of the B. Assoc. for it is a great honour, & as you have so much work to do I am equally glad that you have declined it.[1] I feel, however, convinced that you would have succeeded very well; but if I fancy myself in such a position it actually makes my blood run cold. I look back with amazement at the skill & taste with which the D. of Argyll made a multitude of little speeches at Glasglow. By the way I have not seen the Duke's book,[2] but I formerly thought that some of the articles which appeared in periodicals were very clever, but not very profound. One of these was reviewed in the Saturday Review some years ago; & the fallacy of some main argument was admirably exposed, & I sent the article to you, & you agreed strongly with it.[3] Now I have forgotten this counter-argument & I know I shall be humbugged by the Duke, if I reread him as I suppose I must. There was the other day a rather good review of the Duke's book in the Spectator, & with a new explanation, either by the Duke or Reviewer (I could not make out which) of rudimentary organs; viz that economy of labour & material was a great guiding principle with God (ignoring waste of seed & of young, monsters &c &c), & that making a new plan for the structure of animals was thought & thought was labour, & therefore God kept to a uniform plan & left rudiments. This is no exaggeration. In short, God is a man rather cleverer than us: I wonder they did not suggest that he would suffer from indigestion, if he worked his brains too much.— I am very much obliged for the "Nation" (returned by this post): it is *admirably* good: you say I always guess wrong, but I do not believe anyone, except Asa Gray could have done the thing so well. I would bet even, or 3 to 2, that it is Asa Gray, though one or two passages staggered me.[4]

I finish my Book on "Domestic Animals &c" by a single paragraph answering, or rather throwing doubt, in so far as so little space permits on Asa Gray's doctrine that each variation has been specially ordered or led along a beneficial line. It is foolish to touch such subjects, but there have been so many allusions to what I think about the part which God has played in the formation of organic beings, that I thought it shabby to evade the question. I have even received several letters on subject. One was a funny one from a lady with a whole string of questions, & when I said I could not answer one; she wrote she was perfectly satisfied & it was exactly what she expected.— . . .[5]

Send me a copy of your Insular paper when printed, as several of us want to read it.—

I told Murray not to publish my book blindly, & he has kept the M.S long & is frightened, perhaps with good reason, for I never know when I go too much into detail; & the details are to be printed in smaller type; & at last the M.S is in printer's hands.— In the interval I began a chapter on Man, for which I have long collected materials, but it has grown too long, & I think I shall publish separately a very small volume, "an essay on the origin of mankind:"[6] I have convinced myself of the means by which the Races of man have been mainly formed, but I do not expect I shall convince anyone else.— I wish the dreadful six-month labour of correcting press was over.—

Hensleigh Wedgwood has been very ill, & is sadly pulled down, but is now recovering.—

Give our very kind remembrances to Mrs Hooker & our congratulations on her coming down stairs

Ever yours affect^y^ | C. Darwin

On Feb 13th we go for a week to 6 Queen Anne St.— I wish there was any chance of your being in London & seeing you.—

[1] In the end, Hooker accepted the position of president of the British Association for the Advancement of Science for 1868.

[2] G. D. Campbell (the eighth duke of Argyll), *The reign of law* (London, 1867).

[3] CD's nephew, Henry Parker (1827/8–92), anonymously reviewed an article by Campbell on CD's *Orchids* in *Saturday Review*, 15 November 1862, pp. 589–90.

[4] Anon., 'Popularizing science', *Nation* 5 (1867): 32–4. The author focused on a number of Louis Agassiz's lectures and publications to illustrate the dangers of popularisation.

[5] See letters from M. E. Boole, 13 December 1866 and 17 December [1866], and letter to M. E. Boole, 14 December 1866.

[6] CD eventually used the material in the writing of *Descent* and *Expression.*

To Fritz Müller 22 February [1867]

Down Bromley Kent
Feb 22

My dear Sir . . .

Although you have aided me to so great an extent in many ways, I am going to beg for any information on two other subjects. I am preparing a discussion on "sexual selection", & I want much to know how low down in the animal scale sexual selection of a particular kind extends. Do you know of any lowly organized animals, in which the sexes are separated and in which the male differs from the female in arms of offence, like the horns & tusks of male mammals, or in gaudy plumage & ornaments as with birds & butterflies? I do not refer to secondary sexual characters by which the male is able to discover the female, like the plumed antennæ of Moths, or by which the male is enabled to seize the female, like the curious pincers described by you in some of the lower crustaceans. But what I want to know is how low in the scale sexual differences occur which require some degree of self-consciousness in the males, as weapons by which they fight for the female, or ornaments which attract the opposite sex. Any differences between males & females which follow different habits of life w^{d} have to be excluded. I think you will easily see what I wish to learn. A priori it w^{d} never have been anticipated that insects w^{d} have been attracted by the beautiful colouring of the opposite sex, or by the sounds emitted by the various musical instruments of the male Orthoptera. I know no one so likely to answer this question as yourself. & sh^{d} be grateful for any information however small.

My second subject refers to expression of countenance, to which I have long attended, & in which I feel a keen interest; but to which unfortunately I did not attend when I had the opportunity of observing various races of Man. It has occurred to me that you might without much trouble make a **few** observations for me in the course of some months on Negros, or possibly on native S. Americans; though I care most about Negros. Accordingly I enclose some questions as a guide & if you c^{d} answer me even one or two I sh^{d} feel truly obliged. I am thinking of writing a little essay on the origin of Mankind, as I have been taunted with concealing my opinions; & I sh^{d} do this immediately after the completion of my present book. In this case I sh^{d} add a chapter on the cause or meaning of Expression.

With gratitude for all your great kindness & sincere admiration of all your powers of observation I remain | my dear Sir yours very | sincerely C. Darwin

PS. | You must not give yourself any great trouble about these questions, but possibly you might in the course of a few months be able to observe for me one or two points.

I have sent copies to other quarters of the world
an answer within 6 or 8 months w^{d} be in time.—

If you kept the subject occasionally before your mind, an opportunity of observing some few cases, such for instance as (4) or (5) or (13) &c would almost certainly occur.—

But you must not plague yourself on a subject which will appear trifling to you, but has, I am sure, some considerable interest.

[Enclosure]

Queries about Expression

(1) Is astonishment expressed by the eyes & mouth being opened wide & by the eyebrows being elevated?

(2) Does shame excite a blush, when the colour of the skin allows it to be visible?

(3) When a man is indignant or defiant does he frown, hold his body & head erect, square his shoulders & clench his fists?

(4) When considering deeply on any subject or trying to understand any puzzle does he frown, or wrinkle the skin beneath the lower eyelids?

(5) When in low spirits are the corners of the mouth depressed, & the inner corner or angle of the eybrows raised by that muscle which the French call the 'grief' muscle?

(6) When in good spirits do the eyes sparkle with the skin round & under them a little wrinkled & with the mouth a little extended?

(7) When a man sneers or snarls at another is the corner of the upper lip over the canine teeth raised on the side facing the man whom he addresses?

(8) Can a dogged or obstinate expression be recognized, which is chiefly shewn by the mouth being firmly closed, a lowering brow & slight frown?

(9) Is contempt expressed by a slight protrusion of the lips & turning up of the nose with a slight expiration?

(10) Is disgust shewn by everted lower lip, slightly raised upper lip with sudden expiration something like incipient vomiting?

(11) Is extreme fear expressed in the same general manner as with Europeans?

(12) Is laughter ever carried to such an extreme as to bring tears into the eyes?

(13) When a man wishes to shew that he cannot prevent something being done, or cannot himself do something does he shrug his shoulders, turn inwards his elbows, extend outwards his hands & open the palms?

(14) Do the children when sulky pout, or greatly protrude their lips?

(15) Can guilty or sly or jealous expressions be recognized? tho' I know not how these can be defined.

(16) As a sign to keep silent is a gentle hiss uttered?

(17) Is the head nodded vertically in affirmation, & shaken laterally in negation?

Observations on natives who have had little communications with Europeans would be of course the most valuable, tho' those made on any natives w^{d} be of much interest to me. General remarks on expression are of comparatively little value. A definite description of the countenance under any emotion or frame of mind w^{d} possess much more value, & an answer to any one of the foregoing questions w^{d} be gratefully accepted.

Ch. Darwin
Down Bromley Kent

To Edward Blyth 23 February [1867]

Down. | Bromley. | Kent. S.E.
Feb 23.

My dear Mr Blyth

I have been very much interested by your remarks on the "Origin"; many of which are quite new to me, such as those on mimicry. I knew already some few of the other facts which you mention. I am thinking of writing a short essay on Man & have consequently been much struck with your remarks on the Orang. Do you know C. Vogt's nearly similar remarks on the origin of Man from distinct Ape-families, founded on Gratiolet's observations on the brain? I think you cannot object to my cautiously alluding to your observation on the similarity of the Orang & Malay &c: I think the similarity must be accidental, & I would confirm this by your observation on the S. American genus with respect to the Negro. I do not know what to think about the almost parallel case of Bats. If I had known that you w^{d} have cared for a copy of the new Edit. of the Origin assuredly I w^{d} have sent you one: you will of course receive my book on "Dom. Animals &c" whenever published. I regret much that I did not meet you in London, but during the two last days I was unable to leave the house.

You gave me long since two printed pages Royal 8vo with a black line round the page with notes in very small type, it contains some excellent remarks on sexual plumage of birds evidently by yourself. Please to tell me the title that I may refer to it.

I have picked up more facts on sexual characters, even when you are not discussing the subject from your writings than from those of any one else. Thus in the last N^{o} of "Land & Water" there is a notice which I am sure must be written by you, in which you indicate that the summer plumage of Gulls, Plovers &c is nuptial plumage common to both sexes like that confined to the Drake. This is a new idea to me, if I understand you rightly.— But I presume that you admit that winter plumage, as with the Ptarmigan, may be acquired for a special end.

How I wish that you w^{d} always sign your name to whatever you write. Can you guide me to any papers on sexual differences, especially in colour, in Mammals? I have picked out two cases by you on Bos & Antelopes in the "Indian Field".

Can you tell me whether the canine teeth differ in the sexes in the true Carnivora? But I must not ask any more questions except one:—Do you still maintain the law on sexual plumage in Birds as given in the sheets above referred to? Yarrell gives a rather different law, dependent on there being an annual change in plumage.

But I have written to you at quite unreasonable length— pray forgive me & believe me yours very sincerely— | Ch. Darwin

To A. R. Wallace 23 February 1867

Down, Bromley, Kent, S.E.
February 23, 1867.

Dear Wallace,—

I much regretted that I was unable to call on you, but after Monday I was unable even to leave the house. On Monday evening I called on Bates and put a difficulty before him, which he could not answer, and, as on some former similar occasion, his first suggestion was, "You had better ask Wallace." My difficulty is, why are caterpillars sometimes so beautifully and artistically coloured? Seeing that many are coloured to escape danger, I can hardly attribute their bright colour in other cases to mere physical conditions. Bates says the most gaudy caterpillar he ever saw in Amazonia (of a Sphinx) was conspicuous at the distance of yards from its black and red colouring whilst feeding on large green leaves. If anyone objected to male butterflies having been made beautiful by sexual selection, and asked why should they not have been made beautiful as well as their caterpillars, what would you answer? I could not answer, but should maintain my ground. Will you think over this, and some time, either by letter or when we meet, tell me what you think? Also, I want to know whether your *female* mimetic butterfly is more beautiful and brighter than the male?

When next in London I must get you to show me your Kingfishers.

My health is a dreadful evil; I failed in half my engagements during this last visit to London.— Believe me, yours very sincerely, | C. Darwin.

From A. R. Wallace 24 February [1867]

9, St. Mark's Crescent | N.W.
Feb. 24th.

Dear Darwin

I saw Bates a few days ago & he mentioned to me this difficulty of the catterpillars. I think it is one that can only be solved by special observation. The only probable solution I can imagine is something like this Catterpillars are very similar in form & there are hundreds of species that are only to be distinguished by *colour.*

Now great numbers are protected by their green colours assimilating with foliage or their brown colours resembling bark or twigs. Others are protected by prickles and long hairs—which no doubt render them distasteful to birds, especially to our small birds which I presume are the great destroyers of catterpillars. Now supposing that others, not hairy, are protected by a disagreeable taste or odour, it would be a positive advantage to them never to be mistaken for any of the palatable catterpillars, because a slight wound such as would be caused by a peck of a bird's bill almost always I believe kills a growing catterpillar. Any gaudy & conspicuous colour therefore, that would plainly distinguish them from the brown & green eatable catterpillars, would enable *birds* to recognise them easily as a kind not fit for food, & thus they would escape *seizure* which is as bad as being *eaten.*

Now this can be tested by experiment, by any one who keeps a variety of insectivorous birds. They ought as a rule to refuse to eat and generally refuse to touch gaudy coloured catterpillars, & to devour readily all that have any protective tints. I will ask Mr. Jenner Weir of Blackheath about this, as he has had an aviary for many years & is a very close and acute observer, & I have no doubt will make the experiment this summer.

When our discussion on Mimicry took place a most interesting little fact was mentioned by Mr. Stainton. After *mothing* he is accustomed to throw all the common species to his poultry & once having a lot of young turkeys he threw them a quantity of moths which they eat greedily, but among them was one common *white moth* (*Spilosoma menthastri*) One of the young turkeys took this in his beak, shook his head & threw it down again, another ran to seize it and did the same, and so on, the whole brood in succession rejected it! Mr. Weir tells me that the larva of this moth is hairy & is also rejected by all his birds, which sufficiently accounts for the insect being *very common.* But what is still more curious, another moth much *less common* (*Diaphora mendica*) has the *female* also white, (although the male is quite different) and might at night be easily mistaken for the other! So here we have a case of British mimicry exactly analogous in all its details to that of the *Heliconidæ* & *Danaidæ*; and it is particularly valuable because it is a *direct proof* that Lepidoptera *do* differ in flavour, & that certain flavours *are* distasteful to birds.

My female mimetic butterfly *is* much more beautiful than the male, being metallic blue while the male is dull brown. I sometimes doubt whether sexual selection has acted to produce the colours of *male butterflies.* I have thought that it was merely that it was advantageous for the females to have less brilliant colours, & that colour has been produced merely because in the process of infinite variation *all colours* in turn were produced. Undoubtedly two or three male butterflies do often follow a female, but whether *she* chooses between them or whether the strongest & most active gets her is the question. Cannot this also be decided by experiment? If a lot of common butterflies were bred, say our "brimstone" or better, the "*orange tip*", & the males and females separated & then a certain number of the males *discoloured* by rubbing the wings carefully;—and we were then to turn out a female along with a coloured and a discoloured male into a room or greenhouse, would the female always or in the majority of cases choose the best coloured male.? A series of experiments of this kind carefully carried out would I think settle the question. I will suggest these two classes of experiment at the next meeting of the Entomological & perhaps some country residents may be induced to carry them out. . . .

Yours very faithfully | Alfred R. Wallace

To A. R. Wallace 26 February [1867]

Down. | Bromley. | Kent. S.E.
Feb 26

My dear Wallace

Bates was quite right, you are the man to apply to in a difficulty. I never heard any thing more ingenious than your suggestion & I hope you may be able to prove it true. That is a splendid fact about the white moths: it warms one's very blood to see a theory thus almost proved to be true. With respect to the beauty of male butterflies, I must as yet think that it is due to sexual selection; there is some evidence that Dragon-flies are attracted by bright colours; but what leads me to the above belief is, so many male Orthoptera & Cicadas having musical instruments. This being the case the analogy of birds makes me believe in sexual selection with respect to colour in insects. I wish I had strength & time to make some of the experiments suggested by you; but I thought butterflies w[d] not pair in confinement; I am sure I have heard of some such difficulty.

Many years ago I had a dragon-fly painted with gorgeous colours but I never had an opportunity of fairly trying it.

The reason of my being so much interested just at present about sexual selection is that I have almost resolved to publish a little essay on the Origin of Mankind, & I still strongly think (tho' I failed to convince you, & this to me is the heaviest blow possible) that sexual selection has been the main agent in forming the races of Man.

By the way there is another subject which I shall introduce in my essay, viz expression of countenance; now do you happen to know by any odd chance a very good-natured & acute observer in the Malay Arch. who you think w^{d} make a few easy observations for me on the expression of the Malays when excited by various emotions? For in this case I w^{d} send to such person a list of queries.

I thank you for your *most interesting* letter & remain | yours very sincerely | Ch. Darwin

To Frederic William Farrar 5 March 1867

Down
March 5, 1867

My dear Sir

I am very much obliged for your kind present of your lecture.[1] We have read it aloud with the greatest interest and I agree to every word. I admire your candour and wonderful freedom from prejudice; for I feel an inward conviction that if I had been a great classical scholar I should never have been able to have judged fairly on the subject. As it is, I am one of the root and branch men, and would leave classics to be learnt by those alone who have sufficient zeal and the high taste requisite for their appreciation. You have indeed done a great public service in speaking out so boldly. Scientific men might rail for ever, and it would only be said that they railed at what they did not understand. I was at school at Shrewsbury under a great scholar, D^{r}. Butler I learnt absolutely nothing, except by amusing myself by reading and experimenting in chemistry. D^{r}. Butler somehow found this out and publicly sneered at me before the whole school, for such gross waste of time; I remember he called me a Pococurante,[2] which, not understanding, I thought was a dreadful name. I wish you had shown in your lecture how Science could practically be taught in a great school; I have often heard it objected that this could not be done, and I never knew what to say in answer.

I heartily hope that you may live to see your zeal and labour produce good fruit; and with my best thanks, I remain, my dear Sir | Yours very sincerely | Charles Darwin

[1] F. W. Farrar, 'On some defects in public school education', *Proceedings of the Royal Institution of Great Britain* 5 (1866–9): 26–44.

[2] Pococurante: a habitually uninterested person.

To J. D. Hooker 17 March [1867]

Down.
Mar 17

My dear Hooker …

It is great news about the presidentship; I am very sorry for it, tho' you seem to keep up your spirits.[1] You ask what I have been doing; nothing but blackening proofs with corrections. I do not believe any man in England naturally writes so vile a style as I do. The only fact which I have lately ascertained, & about which I dont know whether you w[d] care, is that a great excess of, or very little pollen produced not the least difference in the average number, weight, or period of germination in the seeds of Ipomœa. I remember saying the contrary to you & M[r] Smith at Kew. But the result is now clear from a great series of trials. On the other hand seeds from this plant, fertilised by pollen from the same flower, weigh less, produce dwarfer plants, but indisputably *germinate quicker* than seeds produced by a cross between two distinct plants.

In your paper on Insular Floras (p. 9) there is what I must think an error, which I before pointed out to you; viz you say that the plants which are wholly distinct from those of nearest continents are often *very common*, instead of very rare.[2] Etty,[3] who has read your paper with great interest, was confounded by this sentence. By the way I have stumbled on two old notes, one that 22 species of European birds occasionally arrive as *chance* wanderer, to the Azores, & secondly the trunks of American trees have been known to be washed on shores of Canary isl[d], by gulf-stream, which returns southward from the Azores. . . .

I have had a very nice letter from Scott at Calcutta: he has been making some good observations on the acclimatisation of seeds from plants of same species, grown in different countries; & likewise on how far European plants will stand the climate of Calcutta; he says he is astonished how well some flourish, & he maintains, if the land were unoccupied, several could easily cross, spreading by seed, the Tropics from N. to South; so he knows how to please me, but I have told him to be cautious, else he will have Dragons down on him.—

I was going to have asked you what sort of a man Benj. Clarke was (I bought his book out of kindness)[4] but I see A. Gray calls him "that ass". He is now going to publish analogous views on Animals, & I have subscribed. He tells me that no single person has or can object to his views on plants; I suspected that perhaps no one noticed them. He tells me a wonderful story of the effects of an inherited mutilation from cutting off the upper half of the ears of wheat for only 3 generations, which I cannot believe; & I have told him no one would believe it, unless he repeats & rerepeats his experiment.—

Farewell my dear old friend | C. Darwin

1 Hooker had accepted the presidency of the British Association for the Advancement of Science for 1868.

2 CD refers to an offprint of Hooker's lecture on insular floras; see also *Journal of Botany* 5 (1867): 23–31.

3 Henrietta Emma Darwin, CD's eldest daughter.

4 Benjamin Clarke, *A new arrangement of phanerogamous plants* (London, 1866).

To Ernst Haeckel 12 April [1867]

Down. | Bromley. | Kent. S.E.
Ap 12.

My dear Sir

I hope you have returned home well in health, & that you have reaped a rich harvest in natural science.[1] I have been intending for some time to write to you about your great work, of which I have lately been reading a good deal.[2] But it makes me almost mad with vexation that I am able to read imperfectly only 2 or 3 pages at a time. The whole book w^{d} be infinitely interesting & useful to me. What has struck me most is the singular clearness with which all the lesser principles & the general philosophy of the subject have been thought out by you & methodically arranged. Your criticism on the struggle for existence offers a good instance how much clearer your thoughts are than mine.

Your whole discussion on dysteologie has struck me as particularly good.[3] But it is hopeless to specify this or that part; the whole seems to me excellent. It is equally hopeless to attempt thanking you for all the honours with which you so repeatedly crown me. I hope that you will not think me impertinent if I make one criticism: some of your remarks on various authors seem to me too severe; but I cannot judge well on this head from being so poor a German scholar. I have however heard complaints from several excellent authorities & admirers of your work on the severity of your criticisms. This seems to me very unfortunate for I have long observed that much severity leads the reader to take the side of the attacked person. I can call to mind distinct instances in which severity produced directly the opposite effect to what was intended. I feel sure that our good friend Huxley, though he has much influence, w[d] have had far more if he had been more moderate & less frequent in his attacks. As you will surely play a great part in science, let me as an older man earnestly beg you to reflect on what I have ventured to say. I know that it is easy to preach & if I had the power of writing with severity I dare say I sh[d] triumph in turning poor devils inside out & exposing all their imbecility. Nevertheless I am convinced that this power does no good, only causes pain. I may add that as we daily see men arriving at opposite conclusions from the same premises it seems to me doubtful policy to speak too positively on any complex subject however much a man may feel convinced of the truth of his own conclusions. Now can you forgive me for my freedom? Though we have met only once I write to you as to an old friend, for I feel thus towards you.

With respect to my own book on Variation under domestication I am making slow, but sure progress in correcting the proofs. I fear that it will interest you but little, & you will be struck how badly I have arranged some of the subjects which you have discussed. The chief use of my book will be in the large accumulation of facts by which certain propositions are I think established. I have indulged in one lengthened hypothesis, but whether this will interest you or any one else, I cannot even conjecture.

I hope before long you will write to me & tell me how you are & what you have been doing & believe me my dear Häckel yours very sincerely | Ch. Darwin

[1] Haeckel had spent from November 1866 to March 1867 travelling and doing research on Tenerife and Lanzarote in the Canary Islands.

[2] Ernst Haeckel, *Generelle Morphologie der Organismen. Allgemeine Grundzüge der organischen Formen-Wissenschaft, mechanisch begründet durch die von Charles Darwin reformirte Descendenz-Theorie* (Berlin, 1866).

[3] Dysteleology: the study of functionless rudimentary organs in animals and plants.

To A. R. Wallace 5 May [1867]

Down. | Bromley. | Kent. S.E.
May 5

My dear Wallace

The offer of your valuable notes is *most* generous, but it w^d vex me to take so much from you, as it is certain that you c^d work up the subject very much better than I could. Therefore I earnestly & without any reservation hope that you will proceed with y^r paper, so that I return y^r notes.

You seem already to have well investigated the subject. I confess on receiving yr note that I felt rather flat at my recent work being almost thrown away, but I did not intend to shew this feeling. As a proof how little advance I had made on the subject, I may mention that though I had been collecting facts on the colouring, & other sexual differences in mammals, your explanation with respect to the females had not occurred to me.[1] I am surprized at my own stupidity, but I have long recognized how much clearer & deeper your insight into matters is than mine. I do not know how far you have attended to the laws of inheritance, so what follows may be obvious to you. I have begun my discussion on sexual selection by shewing that new characters often appear in one sex & are transmitted to that sex alone, & that from some unknown cause such characters apparently appear oftener in the male than in the female. Secondly characters may be developed & be confined to the male, & long afterwards be transferred to the female. 3^{rdly} characters may arise in either sex & be transmitted to both sexes, either in an equal or unequal degree. In this latter case I have supposed that the survival of the fittest has come into play with female birds & kept the female dull-coloured. With respect to the absence of spurs in female gallinaceous birds, I presume that they w^d be in the way during incubation; at least I have got the case of a German breed of fowls in which the hens were spurred, & were found to disturb & break their eggs much.

With respect to the females of deer not having horns, I presume it is to save the loss of organized matter.

In y^r note you speak of sexual selection & protection as sufficient to account for the colouring of all animals, but it seems to me doubtful how far this will come into play with some of the lower animals, such as sea anemones, some corals &c &c—

On the other hand Häckel has recently well shewn that the transparency & absence of colour in the lower oceanic animals, belonging to the most different classes, may be well accounted for on the principle of protection.

Some time or other I sh^{d} like much to know where y^{r} paper on the nests of birds has appeared, & I shall be extremely anxious to read y^{r} paper in the West. Rev.[2] Your paper on the sexual colouring of birds will I have no doubt be very striking.

Forgive me, if you can, for a touch of illiberality about y^{r} paper & believe me yrs very sincerely | Ch. Darwin

[1] Wallace had suggested that the coloration of birds was partly governed by their nesting habits: female (or less frequently, male) birds that sat on exposed nests tended to be protected by being duller in colour than their mates, while both sexes of birds that used concealed or covered nests were brightly coloured, as a result of the unimpeded action of sexual selection.

[2] A. R. Wallace, 'The philosophy of birds' nests', *Intellectual Observer* 11 (1867): 413–20; 'Mimicry and other protective resemblances among animals', *Westminster Review* n.s. 32 (1867): 1–43.

To Charles Kingsley 10 June [1867]

Down. | Bromley. | Kent. S.E.
June 10

My dear Mr Kingsley . . .

I have just finished reading the Duke's book & N. Brit. Rev.;[1] & I sh^{d} very much like for my own sake to make some remarks on them, & as my amanuensis writes so clearly, I hope it will not plague you. The Duke's book strikes me as very well written, very interesting, honest & clever & very arrogant. How coolly he says that even J. S. Mill does not know what he means. Clever as the book is, I think some parts are weak, as about rudimentary organs, & about the diversified structure of humming birds. How strange it is that he sh^{d} freely admit that every detail of structure is of service in the flowers of orchids, & not in the beak of birds. His argument with respect to diversity of structure is much the same as if he were to say that a mechanic w^{d} succeed better in England if he c^{d} do a little work in many trades, than by being a first-rate workman in one trade. I sh^{d} like you to read what I have said upon diversity of structure at 226 in the new Ed. of Origin, which I have ordered to be sent to you. Please also read what I have said

(p. 238) on Beauty.) Other explanations with respect to beauty will no doubt be found out: I think the enclosed ingenious letter by Wallace is worth y^r notice. Is i[t] not absurd to speak of beauty as existing independently of any sentient being to appreciate it? And yet the Duke seems to me thus to speak. With respect to the Deity having created objects beautiful for his own pleasure, I have not a word to say against it but such a view c^d hardly come into a scientific book. In regard to the difference between female birds I believe what you say is very true; & I can shew with fowls that the 2 sexes often vary in correlation. I am glad that you are inclined to admit sexual selection. I have lately been attending much to this subject, & am more than ever convinced of the truth of the view. You will see in the discussion on beauty that I allude to the cause of female birds not being beautiful; but Mr Wallace is going to generalize the same view to a grand extent, for he finds there is almost always a relation between the nature of the nest & the beauty of the female.

No doubt sexual selection seems very improbable when one looks at a peacock's tail, but it is an error to suppose that the female selects each detail of colour. She merely selects beauty, & laws of growth determine the varied zones of colour: thus a circular spot w^d almost certainly become developed into circular zones, in the same manner as I have seen the black wing-bar in pigeons become converted into 3 bars of colour elegantly shaded into each other. The Duke is not quite fair in his attack on me with respect to "correlation of growth"; for I have defined what I mean by it, tho' the term may be a bad one, whilst he uses another definition: "correlation of variation" w^d perhaps have been a better term for me. He depreciates the importance of natural selection, but I presume he w^d not deny that Bakewell, Collins,[2] &c had in one sense made our improved breeds of cattle, yet of course the initial variations have naturally arisen; but until selected, they remained unimportant, & in this same sense natural selection seems to me all-important.

The N. Brit. Rev. seems to me one of the most telling Reviews of the hostile kind, & shews much ability, but not, as you say, much knowledge. The R. lays great stress on our domestic races having been rapidly formed, but I can shew that this is a complete error; it is the work of centuries, probably in some cases of 1000s of years. With respect to the antiquity of the world & the uniformity of its changes, I cannot implicitly believe the mathematicians, seeing what widely different results Haughton Hopkins & Thompson have arrived at. . . .

Is there not great doubt on the bearing of the attraction of gravity with respect to the conservation of energy? The glacial period may make one doubt whether the temperature of the universe is so simple a question. No one can long study the Geolog. work done during the glacial period, & not end profoundly impressed with the necessary lapse of time; & the crust of the earth was at this recent period as thick as now & the force of Nature not more energetic. But what extremely concerns me, is R. statement that I require million of years to make new species; but I have not said so, on contrary, I have lately stated that the change is probably rapid both in formation of single species & of whole groups of species, in comparison with the duration of each species when once formed or in comparison with the time required for the development of a group of species— with respect to Classification, it is the idea of a *natural* classification, which the genealogical explains. The best bit of Review, which c^{d} make me modify wording of few passages in origin is I think about sudden sports, & these I have always thought, but now more clearly see, w^{d} generally be lost by crossing. The R does not however notice, that any variation w^{d}. be more likely to recur in crossed offspring still exposed to same conditions, as those which first caused the parent to vary.— I have moreover expressly stated that I do not believe in the sudden deviation of structure under nature, such as occurs under dom[estication]: but I weakened the sentence in deference to Harvey.—[3]

When speaking of the formation for instance of a new sp. of Bird with long beak Instead of saying, as I have sometimes incautiously done a bird suddenly appeared with a beak particularly longer than that of his fellows, I would now say that of all the birds annually born, some will have a beak a shade longer, & some a shade shorter, & that under conditions or habits of life favouring longer beak, all the individuals, with beaks a little longer would be more apt to survive than those with beaks shorter than average.

The preservation of the longer-beaked birds, would in addition add to the augmented tendency to vary in this same direction.— I have given this idea, but I have not done so in a sufficiently exclusive manner.— . . .

… Pray excuse this unreasonable letter, which you may not think worth the labour of reading; but it has done me good to express my opinion on the 2 works in question, so I hope & think that you will forgive me—

With very sincere thanks for letter believe me my dear Mr Kingsley

yours sincerely | Charles Darwin …

[1] G. D. Campbell (the duke of Argyll), *The reign of law* (London, 1867); [H. C. F. Jenkin], 'The origin of species', *North British Review* 46 (1867): 277–318.

[2] CD refers to Robert or Charles Colling.

[3] In an article in *Gardeners' Chronicle*, 18 February 1860, pp. 145–6, William Henry Harvey had reported the apparent origination of a new species through the abnormal development of *Begonia frigida.*

To Henrietta Emma Darwin 26 July [1867]

July 26th

My dear Etty.—

You are a very good girl to wish for remaining slips of present chapter, but they are enormously altered & 10 folio pages of MS added, & the slips themselves have had to be cut into pieces & rearranged, so I will not send them.

But for the future I shall be only too glad for you to see the slips, as well as Revises.— I will either keep, according to quantity finished, the whole of present chapter till your return, or send part to you.—

All your remarks, criticisms doubts & corrections are excellent, excellent, excellent

Yours affectly. | C. D.

To Charles Kingsley 6 November [1867]

Down. | Bromley. | Kent. S.E.
Nov. 6th

My dear Mr Kingsley

The subject to which you refer is quite new to me & very curious I had no idea that the double function of an excretory passage had even played a part in the history of religion.[1] I agree with what you say on speciality of organs being the best proof of highness in the scale of beings, nevertheless, when man as a standard of comparison is excluded, as with plants, it seems to be nearly impossible to give a good definition of Highness. I do not feel sure that a passage performing a double function, if performed well, ought to be considered a sign of lowness. I suppose that the presence of rudiments must be looked at as an imperfection, but it seems very doubtful whether their records of a former, & in most cases lower, state should be viewed as indices of relative lowness in the scale. Some authors, indeed, have used them as proofs of an opposite position.— It is an extraordinary fact that even man should still bear about his body the plain evidence, as it seems to me, of the former hermaphrodite condition of the parent-form of all the Vertebrata.—

From what you formerly wrote, I had hoped to have seen a review by you on the Reign of Law, but I have not been able to hear of its appearance.

Pray believe me | Yours very sincerely | Charles Darwin

[1] Kingsley had claimed that sex had long been held in contempt by various religions owing to the sharing of urinary and sexual functions by the same organ.

From Charles Kingsley 8 November 1867

Eversley Rectory, | *Winchfield.*
Nov. 8/67.

My dear M^r. Darwin

My thanks for your most interesting letter.

Sex—you will find—plays *the* part in the real ground of all creeds. It is the primæval fact w^h. has to be explained, or misexplained, somehow. I c^d write volumes on this. I may write one little one some day—

As you say—the plain fact that man bears the evidence of a former hermaphrodite type are as indisputable—as they are carefully ignored—

The whole question will have to be reconsidered by us—or by some other wiser race—in the next few Centuries.— *& you will be esteemed then as a prophet.*

Yours ever sincerely | C Kingsley

I have found actually a Darwinian Marchioness!!!!! So even the Swells of the World are beginning to believe in you. The extreme Radical press is staying off from you, because you may be made a Tory & an Aristocrat of. So goes the foolish ignorant world— It will go, believing & disbelieving not according to facts, but to *convenience*. But do you keep yourself—(as you are) "unspotted from the world" as the good book bids all good men do—& then 500 years hence, men will know what you have done for them.

1868

From William Sweetland Dallas 8 January 1868

Yorkshire Philosophical Society | *York*
8 Jan. 1868.

My dear Sir

I this morning sent off the last portion of the Index to your book,[1] & delighted I was to see it slide into the letter box.— I have seen no proofs yet, but I suppose they will come soon,— I have a residue of almost a dozen or a dozen & a half references which got misplaced & which must be inserted on the proofs.— I can only hope that now it is finished you ⟨are⟩ pleased with it & think the delay not wholly thrown away.— I also hope that I may never again let myself in for such another job.— I find the work was just *doubled* by inserting references to all authors' names, as compared with the specimen which I sent you, so that as regards the relation of labour to remuneration I shall look rather blank.— And on the other hand M.r Murray deluded me into ⟨under⟩taking the index by an artful insinuation that it would have the advantage of making me, in a manner, an acquaintance of yours, which, I must say *was* a main cause of my engaging to prepare it, & now, I fear, that acquaintance has been made not in the most favourable manner for me.— I can only repeat that the increased labour spun out the work considerably, so that other & *prior* engagements came in the way after a time, & their mutual hindrance has been most inconvenient for myself.—

However, I must trust to your goodness to excuse whatever may have been due to fault of mine, & at all events not "to hate my name for the rest of your life".—

With all good wishes | I remain | My dear Sir | Yours very truly | W. S. Dallas. . . .

[1] *Variation.*

[George Darwin came second in his year in the mathematical honours examination at the University of Cambridge; the position was known as 'second wrangler'.]

To George Howard Darwin 24 January [1868]

Down. | *Bromley.* | *Kent. S.E.*
Jan 24th.

My dear old fellow

I am so pleased. I congratulate you with all my heart & soul.— I always said from your early days that such energy, perseverance & talent as yours, would be sure to succeed; but I never expected such brilliant success as this.

Again & again I congratulate you. But you have made my hand tremble so I can hardly write. The Telegraph came *here* at eleven.— We have written to W. & Boys.—

God Bless you my dear old fellow— may your life so continue.—

Your affectionate Father | Ch. Darwin

[*The variation of animals and plants under domestication* was published on 30 January 1868.]

To T. H. Huxley 30 January [1868]

Down. | *Bromley.* | *Kent. S.E.*
Jan. 30

My dear Huxley

Most sincere thanks for your kind congratulations. I never received a note from you in my life without pleasure; but whether this will be so after you have read Pangenesis, I am very doubtful. Oh Lord what a blowing up I may receive. I write now partly to say that you must not think of looking at my Book till the summer, when I hope you will read pangenesis, for I care for your opinion on such a subject more than for that of any other man in Europe.— You are so terribly sharp-sighted & so confoundedly honest! But to the day of my death I will always maintain that you have been too sharp-sighted on Hybridism; & the Chapter on the subject in my Book I shd like you to read; not that, as I fear, it will produce any good effect, & be hanged to you.—

I rejoice that your children are all pretty well.— Give Mrs Huxley the enclosed & ask her to look out (for hints) when one of her children is struggling & just going to burst out crying. A dear young lady near here, plagued a very young child for my sake, till it cried, & saw the eyebrows for a second or two beautifully oblique, just before the torrent of tears began.

The sympathy of all our friends about George's success (it is the young Herald)[1] has been a wonderful pleasure to us. George has not slaved himself, which makes his success the more satisfactory.

Farewell my dear Huxley, & do not kill yourself with work | Yours most sincerely | Ch. Darwin

1 George Howard Darwin as a boy had an interest in heraldry.

To Fritz Müller 30 January [1868]

Down Bromley Kent
Jan 30

My dear Sir

I send by this post, by French packet, my new book, the publication of which has been much delayed. The greater part, as you will see, is not meant to be read; but I sh^{d} very much like to hear what you think of "Pangenesis", tho' I fear it will appear to *every one* far too speculative.

I am very much obliged for y^{r} answers, tho' few in number (Oct 5th) about expression.[1] I was especially glad to hear about shrugging the shoulders. You say that an old negro woman, when expressing astonishment, wonderfully resembled a Cebus when astonished; but are you sure that the Cebus opened its mouth?? I *ask* because the Chimpanzee does not open its mouth when astonished or when listening. Please have the kindness to remember that I am very anxious to know whether any monkey when screaming **violently** partially or wholly closes its eyes.

Many thanks for your answers about the Planariæ, & about conspicuous seeds. By the way one of your seeds of the Pavonia has germinated. I sent the Solanum-like flower to Hooker, & I believe he has written to you. How strange it is that the same anomalous reduction of one leaf sh^{d} occur in several families! Gesneria pendulina (this is the sp. name) is certainly not dimorphic. Your Plumbago is Zeylanica an Indian species, I suppose naturalized with you; it shed its first flower-buds but it is now producing others, & I think it will turn out dimorphic. The flowers of Escholzia when crossed with pollen from a distinct plant produced 91 per cent of capsules; when self-fertilized the flowers produced only 66 per cent of capsules. An equal number of crossed & self-fertilized capsules contained seed by weight in the proportion of 100 to 71.

Nevertheless the self-fert. flowers produced an abundance of seed. I enclose a few crossed seeds in hopes that you will raise a plant, cover it with a net, & observe whether it is self-fertile; at the same time allowing several uncovered plants to produce capsules; for the sterility formerly observed by you seems to me very curious.

With sincere thanks for your never failing kindness believe me my dear Sir | yours very faithfully | Charles Darwin

You will find your most valuable observations on self-sterile Orchids given in the second volume.—

[1] Müller's answers to CD queries about expression have not been found; for the queries, see the letter to Fritz Müller, 22 February [1867].

[The following letter has been translated from the original German.]

From Ernst Haeckel [before 6 February 1868]

Dear, most esteemed Sir!

For many weeks now ⟨I have⟩ wanted to write to you almost every day but an unusual number of varied tasks and interruptions have prevented me from doing so until now. I don't really need to assure you that I nevertheless thought of you daily. For aside from the feelings of highest personal regard in which I hold you, my daily work would also remind me of you every day. This winter I again held a lecture series on "Darwin's theory of evolution which is very well received. ⟨This is the⟩ best attended course of ⟨lectures⟩ and I have about 200 ⟨ ⟩ students from all faculties, ⟨ ⟩, teachers, agriculturalists etc. ⟨I⟩ have someone transcribe the lectures, which are kept on a popular level, and next summer they will appear in print.

First of all, allow me to express my most cordial thanks to you for kindly sending me your book "on the variation of animals and plants". Your rich and multifaceted knowledge in all branches of biology and especially in the natural history of domestic animals and cultivated plants has amazed me anew, and I was most keenly interested in many of the facts you mentioned that are new and unknown to me. Still, all the special proofs that you produce for the theory of selectio naturalis do not hold the same kind of interest for me as they will for most readers. I was already so completely convinced of the truth of "natural selection" by your first work on the origin of species that all such special proofs are only of secondary interest. Many German natural scientists still take the view that your selection theory must be proved through numerous individual arguments, and for these your present book will be more important than the first. For me the opposite is the case; although of course I prize very highly all the further special proofs that you produce, if only because of these others.

But I do believe that every scientist who possesses philosophical understanding must already recognise the truth of natural selection from the three undeniable facts of 1) heredity, 2) adaptation, 3) struggle for existence. But most understand nothing of these simple premises and just as little of embryology and palaeontology, and so it's hardly surprising then that they still stick to their foolish opposition.

Jena is now the centre of German Darwinism. Apart from me and Gegenbaur (who is now editing his entire "Comparative Anatomy" in the second revised "phylogenetic" edition) there are another two students here who are working very eagerly for your theory.

My assistant and ⟨companion on⟩ the Canaries trip, ⟨ ⟩ talented young Russian ⟨ ⟩ *Miklucho*, is doing excellent work on the phylogeny of vertebrates. He made the very fine discovery on Lanzarote that the *Selachiae* possess a rudimentary *swim-bladder*, which is very important for my genealogy of vertebrates and for the fact that the character of the lungs emerging from the swim-bladder was already present in the Selachiae as the ancestor of the Amphirhinae. Huxley will send you a copy of his paper. I enclose a photograph of me and Miklucho in our Canaries travelling gear as "Medusae fishermen.

Another keen Darwinist in Jena is Dr *Dohrn*, also a pupil of Gegenbaur and myself. He is very busy ⟨with the⟩ anatomy of the anthropods. ⟨However,⟩ I fear that he proceeds too ⟨ ⟩ with it. His ⟨　　⟩ theory is not tempered enough by his judgment and I think that ⟨　　⟩ his comparative-anatomical correlations are very risky.

You will have received a dissertation "Über den Ursprung der Sprache" from my cousin *Wilhelm Bleek*[1] in Cape Town, who has been studying the language of the Bushmen, Hottentots and Kaffirs there for 13 years. He was earlier a theologian, and is now a Darwinist. At the time he also defended Bishop Colenso.

I still have your plant geography of the Canary Islands by Webb and Berthelot.[2] Since it is useful for my elaboration of the Canaries trip, could I ask you to let me keep it for ⟨another few⟩ months, providing you do not need it yourself, in which case I will return it at once.

My main task at the moment is the finer anatomy and developmental history of the siphonophores, which is very interesting but also very difficult. The polymorphism of these animals and their reduction to simple medusae and polyps provide excellent proofs of the theory of descent.

Otherwise I am very well and live very happily with my young wife[3] in the quiet seclusion of Jena. Life is as simple and quiet here as in the country. We do not yet even have a railway, which would link us more closely to Weimar and thus with the world of culture. But this seclusion, as you know from (Bromley) *Down*, has great advantages!

⟨　　⟩ autumn, shortly after my wedding (on 20 August) I went on a very nice trip with my wife through the Bavarian, Tyrolian and Swiss Alps. However, I nearly perished when I climbed an 8,000 foot high rock face (the Tristenspitze in North Tyrol) and could move neither forwards nor backwards for 3 hours. Even now I can scarcely comprehend how I got back down alive.

Hopefully I will soon hear some good news about your health. Please give my most respectful regards to your highly honoured wife and Miss Darwin.

With continuing regard | yours truly | Ernst Haeckel

1 W. H. I. Bleek, *Über den Ursprung der Sprache* (On the origin of language), edited by Ernst Haeckel (Weimar, 1868).

2 Part of P. B. Webb and Sabin Berthelot, 1836–50. *Histoire naturelle des Iles Canaries* (Paris, 1836–50).

3 Agnes Haeckel.

To J. D. Hooker 10 February [1868]

Down
Feb. 10th

My dear Hooker

What is the good of having a friend, if one may not boast to him? I heard yesterday that Murray has sold in a week the whole edit. of 1500 copies of my book, & the sale so pressing that he has agreed with Clowes to get another edit in 14 days! This has done me a world of good for I had got into a sort of dogged hatred of my book. And now there has appeared a review in the Pall Mall which has pleased me excessively, more perhaps than is reasonable.

I am quite content & do not now care how much I may be pitched into.— If by any chance you sh[d]. hear who wrote the article in the Pall Mall, do please tell me: it is some one who writes capitally & knows the subject.

I went to luncheon on Sunday to Lubbocks,[1] partly in hopes of seeing you, & be hanged to you, you were not there.

Your cock-a-hoop Friend | C. D

[1] John Lubbock.

To B. D. Walsh 14 February 1868

Down. | Bromley. | Kent. S.E.
Feb. 14. 1868

My dear Sir,

I want you to ransack your memory or notebook for some scraps of information, and I know you will excuse me troubling you. I am working on Sexual selection and all sorts of facts would be most useful tome.

I will specify some points. In the N. American Homoptera and Orthoptera do the males and females ever differ conspicuously in colour?

Can you give me any proofs of the females being attracted by the music of the males?

I want to hear anything about the battles of male insects, Any facts about the attachment or love of the male to the female or reciprocally.—

Have you any idea of the use of the strange horns in the male Lamellicorns?

I especially want to hear of insects in which the males and females are very unequal in numbers; especially of the apparently rarer cases of the females being in excess. I want to see what bearing the relative inequality of number has on Sexual Selection. You who so thoroughly understand my views will easily understand the class of facts I want; I shall be grateful if you will ransack your memory. Do your Sibellulæ differ much in colour sexually?

I wrote a little time ago to you saying that my new book had been despatched to New York— It has had a very large sale here.

I hope the world progresses favourably in all ways with you.—

Believe me | Yours very sincerely | Charles Darwin

To J. D. Hooker 23 February [1868]

Down.
Feb. 23[d]

My dear Hooker

I have had almost as many letters to write of late, as you can have—viz from 8–10 per diem,—chiefly getting up facts on sexual selection; therefore I have felt no inclination to write to you, & now I mean to write solely about my Book, for my own satisfaction, & not at all for yours.— The first Edit was 1500 copies, & now the second is printed off,—sharp work. Did you look at the Review in the Athenæum, showing profound contempt of me. I feel convinced it is by Owen— Pouchet—M[r] **Charles** Darwin, always in full & other little strokes.[1] It is a shame that he sh[d] have said that I have taken much from Pouchet without acknowledgement; for I took literally nothing, there being nothing to take. There is capital R. in Gardeners Chronicle—which will sell the book, if anything will— I do not quite see whether I or the writer is in a muddle about man *causing* variability. If a man drops a bit of iron into Sulphuric acid he does not cause the affinities to come into play, yet he may be said to make Sulphate of Iron. I do not know how to avoid ambiguity. After what Pall Mall & G. Chronicle have said, I do not care a d—. for Owen.—

I fear Pangenesis is still-born. Bates says he has read it twice & is not sure that he understands it.— H. Spencer says the view is quite different from his (& this a great relief to me, as I feared to be accused of plagiarism, but utterly failed to be sure what he meant, so thought it safest to give my view as almost same as his)

& he says he is not sure he understands it. Sir J. Lubbock says he shall wait, before he expresses his opinion, & see what the Reviews say. Am I not a poor Devil; yet I took such pains, I must think that I expressed myself clearly Old Sir H. Holland says he has read it twice & thinks it very tough, but believes that sooner or later "some view akin to it" will be accepted.

You will think me very self-sufficient, when I declare that I feel *sure* if Pangenesis is now still born it will thank God at some future time reappear, begotten by some other Father, & christened by some other name.—

Have you ever met with any tangible & clear view of what takes place in generation, whether by seeds or buds.— Or how a long-lost character can possibly reappear—or how the male element can possibly affect the mother-plant—or the mother animal so that her future progeny are affected. Now all these points & many others are connected to-gether,—whether truly or falsely is another question—by Pangenesis.— You see I die hard, & stick up for my poor child.—

This letter is written for my own satisfaction & not for yours—so bear it.

Yours affectionately | Ch. Darwin

[1] The review of *Variation* in the *Athenæum*, 15 February 1868, pp. 243–4, was by John Robertson.

To John Jenner Weir [6 March 1868]

6, Queen St. | Cavendish Sq^{e} | W.

My dear Sir

I have come here for a few weeks for a little change & rest. Just as I was leaving home I received your first note & yesterday a second; & both are most interesting & valuable to me.—

That is a very curious observation about the Gold-finches beak, but one would hardly like to trust it without measurement or comparison of the beaks of several male & female birds; for I do not understand that you yourself assert that the beak of the male is sensibly longer than that of the female.— If you come across any acute bird-catchers (I do **not** mean to ask you to go after them); I wish you would ask what is their impression on the relative numbers of the sexes of any birds, which they habitually catch, & whether some years males are more numerous & some years females.—

I see that I must trust to analogy, (an unsafe support) for sexual selection in regard to colour in Butterflies. You speak of the Brimstone Butterfly & genus Edusa (I forget what this is, & have no books here, unless it it is Colias) not opening their wings; in one of my notes to M^r. Stainton I asked him, (but he could or did not answer) whether Butterflies such as the Fritillaries with wings bright beneath & above, opened & shut their wings more than Vanessæ most of which I think are obscure on the under surface.

That is a most curious observation about the red underwing moth & the Robin; & strongly supports a suggestion, (which I thought hardly credible) of A. R. Wallace viz that the immense wings of some exotic Lepidoptera served as a protection from difficulty of Birds seizing them. I will probably quote your case.—

No doubt Dr Hooker collected the Kerguelen moth, for I remember he told me of the case, when I suggested in the Origin the explanation of the Coleoptera of Madeira being apterous; but he did not know what had become of the specimens.—

I am quite delighted to hear that you are observing coloured Birds, though the probability, I suppose, will be that no sure result will be gained.— I am accustomed with my numerous experiments with plants to be well satisfied if I get any good result in one case out of five.—

With most sincere thanks for your great kindness— | My dear Sir | Yours very sincerely | Ch. Darwin

You will not be able to read all my book—too much detail—some of chaptr in 2^d Vol. are curious I think.—

If any man wants to gain a good opinion of his fellow men, he ought to do what I am doing pester them with letters.—

From J. D. Hooker 20 May 1868

Kew
May 20th/68.

Dear old Darwin

What an age it is since we have corresponded,—I hope I have written since my little trip to Wales with Huxley, which was *perfect*

— Since then I have been for 3 days to see my sister in Torquay,[1] & nowhere else. at Torquay I had a good lecture on "Kent's hole" from Pengelly, who does it thoroughly well.

My time has been actively employed in garden duties, out of doors & indoors, with an occasional diversion at your volumes & Lyell's last[2] You greatly underrate the interest of your's, it is capital reading, putting aside all question of it's matter, which will, if foreigners deign to read it at all, do you more credit in their eyes than all your other works put together. (I have not read $\frac{1}{4}$ of it). Bentham has, & now I think, unreservedly, acknowledged himself a convert to Darwinism! this I quite expected, would be the case with many: a few will still hold back & flaunt the "rag of protection" till your next part appears, holding that cultivation is no argument,—when,—the said rag, being worn back to the rope, & no longer visible, they will gracefully haul it down—

It is so long since we have corresponded, that I do not know what the deuce to write about!— We are all pretty well: my wife expects her confinement in 10 days, & is as miserable as usual, with Heart-burn, dyspepsia, palpitation & every imaginable minor evil of the coming event. Charlie has had measles lightly, at school— The Governess & children go to Eastbourne in a day or two.

Andrew Murrays 2^{d} & 3^{d} parts are better than the first. How do you like Wallace's paper?[3] the more I read the more struck I am with the *great* ability of the man.

I have finished the Reign of Law[4] with utter disgust—& uncontrollable indignation;—considering his birth education & position, I regard him as lower than Owen— his suppressed sneers at you are of a far lower order of sneer than the malicious sneers of Owen. I like a man to sneer at me out of malice & envy—but can not stand a man's sneering at me from a top of a high Horse— The preliminary reasoning on the principles of flight appear to me radically unsound. The idea of God being compelled to dab in rudimentary organs *to keep up appearances*! as it were, is very droll. The little man writes extremely well, & expresses himself with admirable facility— in fact he has a fatal facility for handling things he does not fully understand, & which he has not the time, & probably not the power to grasp the principles of.

Lyell's vol II. is I think a wonderful book, better than all subsequent Editions to Ed. 1. put together— What do you think of it? I have not had time to read all of it—yet. . . .

I am used up, & have nothing more to say— I feel my barrenness of scientific matter to communicate creeping over me every day now—& the tide of scientific literature is already up to my knees. The time was when I had now & then something to communicate that you cared to know— that is all changed now, & I feel very low at times about it.— I begin to despair of doing anything—even at Insular Flora again, wherein I see that I could still do much. perhaps when this Norwich meeting is over I shall feel more at ease. I would give 100 gs. that it were over, even with a failure a fiasco or worse. The address is *nowhere* yet & I look on its prospect with a loathing that cannot be uttered.[5] Tomorrow I go to see Fergusson to *encourage* him about his prospective Lecture at the meeting!— God pity us both—the "blind leading the blind"— I shall have to play the hypocrite with a vengeance—

Ever yr affec | J D Hooker

1 Elizabeth Evans-Lombe.

2 Hooker refers to *Variation* and to the 10th edition of Charles Lyell's *Principles of geology* (London, 1867–8).

3 The second part of the first (and only) volume of Andrew Murray's *Journal of Travel and Natural History* contained an article by Alfred Russel Wallace on birds' nests.

4 G. D. Campbell (the eighth duke of Argyll), *The reign of law* (London, 1867).

5 Hooker was to be president at the British Association for the Advancement of Science's annual meeting in Norwich in August 1868.

To J. D. Hooker 21 May [1868]

Down—
May 21st

My dear Hooker

I know that you have been overworking yourself, & that makes you think that you are doing nothing in science. If this is the case (which I do not believe) your intellect has all run to letter-writing, for I never in all my life received a pleasanter one than your last.— It greatly amused us all.

How dreadfully severe you are on the little Duke;—I really think too severe; but then I am no fair judge, for a Duke in my eyes is no common mortal, & not to be judged by common rules! I pity you from the bottom of my soul about the Address— it makes my flesh creep; but when I pitied you to Huxley, he w^{d} not join at all, & would only say that you did & delivered your Insular-Flora-Lecture so admirably in every way that he would not bestow any pity on you.— He felt certain that you would keep your head high up.— Nevertheless I wish to God it was all over for your sake.— I think from several long talks that Huxley will give an *excellent* & *original* Lecture on Geograph. Distrib. of Birds. I have been working very hard, too hard, of late on Sexual Selection, which turns out a gigantic subject, & almost every day new subjects turn up, requiring investigation & leading to endless letters & searches through books. I am bothered, also, with heaps of foolish letters on all sorts of subjects. But I am much interested in my subject & sometimes see gleams of light.— All my other letters have prevented me indulging myself in writing to you; but I suddenly found the Locust-grass yesterday in flower & had to despatch it at once.[1] I suppose some of your assistants will be able to make the genus out, without great trouble. I have done little in experiments of late; but I find that mignonette is absolutely sterile with pollen from same plant. Anyone who saw stamen after stamen bending upwards & shedding pollen over the stigmas of the same flower would declare that the structure was an admirable contrivance for self-fertilisation.— How utterly mysterious it is that there sh^{d} be some difference in ovules & contents of pollen-grains (for the tubes penetrate own stigma) causing fertilisation when these are taken from *any* two distinct plants, & invariably leading to impotence when taken from the same plant! By Jove even Pan. won't explain this. It is a comfort to me to think that you will be surely haunted on your death-bed for not honouring the great God Pan.— I am quite delighted at what you say about my book & about Bentham. When writing it, I was much interested in some parts, but latterly I thought quite as poorly of it, as even the Athenæum. It ought to be read abroad for the sake of the booksellers; for five editions have come or are coming out abroad!— I am ashamed to say that I have read only the organic part of Lyell, & I admire all that I have read as much as you. It is a comfort to know that *possibly* when one is 70 years old one's brain may be good for work.

It drives me mad & I know it does you too, that one has no time for reading anything beyond what must be read: my room is encumbered with unread books.—

I agree about Wallace's wonderful cleverness; but he is not cautious enough in my opinion. I find I must (& I always distrust myself when I differ from him) separate rather widely from him all about Bird's nests & protection; he is riding that hobby to death.— I never read anything so miserable as Andrew Murray's criticism on Wallace in the last no^{r} of his Journal: I believe this Journal will die & I shall not cry: what a contrast with the old Nat. Hist. Review.— I am very sorry to hear how uncomfortable Mrs Hooker is.— poor women— good Lord how wretched my wife used to be

Farewell— it has done me good scrawling to you— Your's affectionately | C. Darwin

[1] The grass was grown from seeds found in locust dung sent to CD from South Africa.

To J. D. Hooker 17 [June 1868]

Down. | Bromley. | Kent. S.E.
17th

My dear Hooker

What a man you are for sympathy. Your note has pleased us all greatly.— I was made "Eques" some months ago, but did not think much about it. Now by Jove we all do; but you in fact have knighted me![1] I am glad you were at the Messiah: it is the one thing that I sh^{d} like to hear again, but I daresay I sh^{d} find my soul too dried up to appreciate it, as in old days; & then I sh^{d} feel very flat, for it is a horrid bore to feel, as I constantly do, that I am a withered leaf for every subject except science. It sometimes makes me hate science, though God knows I ought to be thankful for such a perennial interest which makes me forget for some hours every day my accursed stomach.—

Frank was at the Messiah & he brought out, on coming home, such a string of "tremendous", "awful", "awfully grand" &c &c.— By the way Frank is beginning in some earnest Botany, & dissecting ovules &c & bewailing the difficulty of anatropous &c— primines, secundines &c— it drives him half mad, & I cannot help him—

I wish you had told me how you get on with your address; I am fearfully interested on this head. I fear that there is no possibility of your coming here for a Sunday; but it is my firm conviction that the Down air would inspire you with some grand original thoughts & sublime sentences.

By the way I do not agree about "The Trumpet shall sound"; for I remember being struck with surprise & then with admiration at there *not* being at the words an overpowering blast of the Trumpet.

I have received the Duke of A. It is good news that Mrs. Hooker is recovering so well.

Yours affectionately | C. Darwin

[1] CD refers to his receiving the Prussian Order of Merit.

[CD and his family rented a cottage at Freshwater on the Isle of Wight from Julia Margaret Cameron, the photographer, from 17 July to 20 August 1868.]

From Julia Margaret Cameron [before 10 July 1868]

My dear Mr. Darwin

My Bromley people rejoice for us in your coming to be our Neighbors & friends as we hope—

A hurried answer I send to your note—

We have in occasional use a person of the highest respectability & usefulness who acts as House Maid or House Keeper a general help— We give her a shilling a day & food— We could ask her to keep disengaged for you to try her if you want help— We have said nothing abt. our gardener our last Tenant Lady Manners paid 14/ a week to our gardener but if you like to have our gardener for half the time & pay 7/ we will pay the other 7/s thus yr. garden & lawn will be his care

To day is a great & busy day in this place yet I write this word rather than delay yr. answer tho I was up till past midnight at work & still have much to do to be ready to open our first Reading Room at Fresh Water at 1 PM

Yours truly | Julia M. Cameron

Our Tenants bring their own Plate & Linen

To J. D. Hooker 14 July 1868

Down. | *Bromley.* | *Kent. S.E.*
July 14— | 1868

My dear Hooker

It will be the most delightful thing in the world if you can pay us a visit at Freshwater.

We were to have started on Thursday, but I have been so baddish of late that I am doubtful whether the journey w^{d} be endurable. As soon as we are settled & I feel that I have even moderate powers of talk in me I will at once let you hear. I am glad you are going to say in your address that you have not time to work up an elaborate affair, for I think every one will see that this is reasonable in the head-man of a large establishment. It cuts me to the quick to be honest, but I think you w^{d} be wise not to touch on Pangenesis. It has so very [few] friends; Bentham as you know, is very doubtful or hostile; Victor Carus dead against it; & Alf. de Candolle says he likes it much the least of the whole book. By the way I was much pleased & surprized by a very long letter from Decandolle, in which he shews he is fond of speculation. Altho' I advise thus about Pan. my conviction is unshaken that it will hereafter be looked at as the best hypothesis of generation, inheritance & development.

But I must write no more so Goodbye—

We are very glad to hear so capital an acc^{t} of Mrs Hooker & the baby.

yours affectionately | Ch Darwin

Perhaps you mean to cut up Pangenesis— if so, I have not a word to say in opposition

To J. D. Hooker 28 July [1868]

Dumbola Lodge | Freshwater | I. of Wight
July 28th

My dear Hooker

We are very sincerely sorry to hear about your little girl. It is miserable for you, but I hope the poor little thing does not suffer much beyond exhaustion, & I have always thought that very young children do not suffer like older ones. I had not heard of the prevalence of infantile diarrhœa, but it is not surprising under such extraordinary weather.

Your work for B. Assoc. must now be extra repulsive to you. I am glad to hear that you are going to touch on the statement that the belief in Nat. selection is passing away; I do not suppose that even the Athenæum w^d. pretend that the belief in the common descent of species is passing away, & this is the more important point. This now *almost universal* belief in the evolution (somehow) of species I think may be fairly attributed in large part to the "Origin." It w^d be well for you to look at short Introduction of Owen's Anat. of Invertebrata, & see how fully he admits the descent of species.[1]

Of Origin, 4 English editions, 1 or 2 American; 2 French, 2 German, 1 Dutch,—1 Italian & several (as I was told) Russian, editions. The translations of my Book on Var. under Domestication are the result of the Origin; & of these 2. English 1. American, 1 German, 1 French, 1 Italian & 1. Russian, have appeared or will soon appear.—

Ernst Häckel wrote to me a week or two ago that new discussions & Reviews of the Origin are continually still coming out in Germany, where the interest on subject certainly does not diminish. I have seen some of these discussions & they are good ones.— I apprehend that the interest on subject has not died out in N. America, from observing in Prof. & M^rs. Agassizs Book on Brazil how excessively anxious he is to destroy me.—[2] In regard to this country, everyone can judge for himself; but you would not say interest was dying out, if you were to look at last no^r of the Anthropological Review, in which I am incessantly sneered at. I think Lyell's Principles will produce considerable effect.[3]

I hope I have given you the sort of information which you want. My head is rather unsteady which makes my hand-writing worse. than usual.— Please keep the Books for me.— We shall be very anxious to hear about your poor Baby.

My dear old Friend | Yours affect. | C. Darwin

If you agree about the non-acceptance of nat. selection, it seems to me a very striking fact that the Newtonian theory of gravitation, which seems to ev[er]yone now so certain & plain, was rejected by a man so extraordinary able as Leibnitz. The truth will not penetrate a preoccupied mind.

Wallace in Westminster Review in article on Protection has good passage, contrasting the success of Natural Selection, & its gro[w]th with the comprehension of new classes of facts, with false theories, such as the Quinarian Theory & that of Polarity by poor Forbes, both of which were promulgated with high advantages, & the first temporarily accepted.—[4]

[1] Richard Owen, *Anatomy of vertebrates* (London, 1868).
[2] Louis and Elizabeth Agassiz, *A journey in Brazil* (Boston, 1868).
[3] Charles Lyell, *Principles of geology*, 10th edition (London, 1867–8).
[4] [A. R. Wallace], 'Mimicry and other protective resemblances among animals', *Westminster Review* n.s. 32 (1867): 1–43. CD refers to Edward Forbes.

[Hooker delivered the presidential address to the annual meeting of the British Association for the Advancement of Science at Norwich on 19 August 1868; he devoted a considerable portion of his address to CD's scientific work and his theory of natural selection.]

To J. D. Hooker 23 August [1868]

Down. | Bromley. | Kent. S.E.
Sunday Aug 23

(Read this when at Kew.)
My dear old Friend.—

I have received your note. I can hardly say how pleased I have been at the success of your address & of the whole Meeting— I have seen the Times, Telegraph, Spectator & Athenæum; & have heard of other favourable newspapers & have ordered a bundle. There is a chorus of praise. The Times reported miserably, ie as far as errata were concerned, but I was very glad at the Leader, for I thought the way you brought in the megalithic monuments most happy. … The Spectator pitches a little into you about Theology, in accordance with its usual spirit; for there is some writer in the Spectator who is the most ardent admirer of the Duke of Argyll. Your great success has rejoiced my heart. I have just carefully read the whole Address in the Athenæum; & though, as you know, I liked it very much when you read it to me; yet as I was trying all the time to find fault, I missed to a certain extent the effect as a whole; & this now appears to me most striking & excellent. How you must rejoice at all your bothering labour & anxiety having had so grand an end. I must say a word about myself: never has such a Eulogium been passed on me & it makes me very proud. I cannot get over my *amazement* at what you say about my Botanical work. By Jove, as far as my memory goes, you have strengthened instead of weakened some of the expressions. What is far more important, than anything personal, is the conviction which I feel that you will have immensely advanced the belief in the evolution of species. This will follow from the publicity of the occasion, your position, so responsible, as President, & your own high reputation. It will make a great step in public opinion I feel sure, & I had not thought of this before.— The Athenæum takes your snubbing with the utmost mildness.— I certainly do rejoice over this snubbing, & hope Owen will feel it a little.— — Whenever you have *spare* time to write again, tell me, whether any Astronomers took your remarks in ill part: as they now stand they do not seem at all too harsh or presumptuous. Many of your sentences strike me as extremely felicitous & eloquent. That of Lyell's "underpinning" is capital. Tell me was Lyell pleased: I am so glad that you remembered my old Dedication. Was Wallace pleased?

How about Photographs? Can you spare time for a line to our dear Mrs. Cameron. She came to see us off & loaded us with presents of Photographs, & Erasmus called after her "Mrs Cameron there are six people in this house all in love with you". When I paid her: she cried out "oh what a lot of money" & ran to boast to her husband!!

Tennyson talked of you in a most friendly way, & took all your snubbing most amiably.—

I must not write any more, though I am in tremendous spirits at your brilliant success.—

Yours ever Affect | C. Darwin

To Gaston de Saporta 24 September 1868

Down. Bromley. Kent. S.E.
Sep: 24 1868.

Dear Sir.

Owing to your letter of Sep: 6 having been addressed to London and not at once fowarded here, I received it only 2 or 3 days ago, otherwise I should not have allowed so long an interval to have elapsed before thanking you most sincerely for the honour which you have done me— Your letter *abounds* with statements & remarks of the highest interest to me. A few years ago it would have been thought quite incredible that the Genus Magnolia should have existed so long ago; & how much light the Antiquity of the genus Fagus throws on the distribution of the existing species on which subject, I have often felt much surprise. I fully appreciate the importance of your observations on the antiquity of certain varieties— It is also surprising to find that paleontology aids so much in giving us the origin of our fruit trees.— I am particularly obliged to you for telling me of the excellent instance of the direct action of pollen in Pistacia— As I have formerly read with great interest many of your papers on fossil plants you may believe with what high satisfaction I hear that you are a believer in the gradual evolution of Species— I had supposed that my book on the origin of species, had made very little impression in France and therefore it delights me to hear a different statement from you. All the great authorities of the Institute seem firmly resolved to believe in the immutability of species, and this has always astonished me in the country which has given birth to Buffon, Lamarck & Geoffroy S[t] Hilaire. Almost the one exception, as far as I know is M[r]. Gaudry—& I think he will be soon one of the Chief Leaders in Zoological Paleontology in Europe & now I am delighted to hear that in the sister department of Botany you take nearly the same view.

With *cordial* thanks & the most sincere respect | I beg leave to remain dear Sir | Yours faithfully & obliged | Charles Darwin

[Adam Sedgwick, Woodwardian Professor of geology at Cambridge University and fellow of Trinity College, had made a geological tour of North Wales with CD in 1831; the trip made a lasting impression on CD. Sedgwick was also a prebendary of Norwich Cathedral. George Howard Darwin had recently been admitted to a fellowship at Trinity College, Cambridge]

From Adam Sedgwick 11 October 1868

Cambridge
Oct 11 | 1868

My dear Darwin

I returned to College (from my Cathedral residence at Norwich) in time to assist at the *admission* of the new Fellows. I shook hands with them in our ante chapel, not knowing the name of any one of them; & little suspecting that your son was one of them. I heard their names afterwards & greatly rejoiced I was to find that the one who is junior on the list was your living representation, I tried to find him out, but he was gone; as I was told by a gyp.

Let me send you my warmest congratulations And now I do hope that you will again sometimes come among us— All my old friends are dead, or have left the University— So that here I am living in solitude, for I cannot bear to go out to any parties— Considering my great age—for I am far advanced in the 84th. year—I am in many respects a strong old fellow. Yet my organic machinery is sadly out of tune— —E.G. I have in my heart an indurated mitral valve, which makes all climbing impossible. 2. I have an enlarged prostate gland, which makes me unfit for society. (3) I have a stone in my right kidney & 4thly. I am liable to dangerous attacks of giddiness, which sometimes fell me to the ground—as if I were shot thro' the head— But these maladies (if I may so call them) have not increased during the last two or three years: & after several irritating & alarming struggles I have settled down into a kind new equilibrium among my animal functions— Do come down with a family party & renew your acquaintance with Cambridge, & the few old friends left in it— New friends you would find in plenty—

There! I have been slopping my ink so 'tis time for me to conclude. My eyes do not—without a higher power than I am using—allow me to read what I am writing

Again accept my congratulations & believe me while my heart beats & my lungs heave Ever yours in all christian truth and good will | A Sedgwick

To Adam Sedgwick 13 October 1868

Down. | Bromley. | Kent. S.E.
Oct 13 1868

My dear Professor Sedgwick

I have been deeply gratified by your kindness in remembering me & sending your congratulations on my son's election as a Fellow of Trinity. This certainly is a very great honour to him & a source of much pleasure to me. I often think of the Geological tour in N. Wales, when you allowed me to accompany you, & which was so great a pleasure & advantage to me. You, also, during my expedition in the Beagle sent me several encouraging messages, the actual words of some of which I still remember.[1] With these recollections & feelings you will believe how deeply I have been gratified by receiving your letter.

I am very sorry to hear that you suffer from so many serious complaints; but considering how often you were formerly unwell you seem to me a wonderful instance of prolonged vigour of mind & body. As for myself I am always an invalid, & fear that I shall never see Cambridge again, though from the many happy days spent there I sh[d] much like to do so.

With the most sincere respect pray believe me | your obliged friend | Charles Darwin

[1] Sedgwick's letters to CD while he was on HMS *Beagle* have not been found; however, CD's sister Susan sent CD an extract from a letter from Sedgwick to Samuel Butler: 'He is doing admirably in S. America, & has already sent home a Collection above all praise.— It was the best thing in the world for him that he went out on the Voyage of Discovery— There was some risk of his turning out an idle man: but his character will now be fixed, & if God spare his life, he will have a great name among the Naturalists of Europe.'

To W. D. Fox 21 October [1868]

Down. | Bromley. | Kent. S.E.
Oct. 21

My dear Fox

I am rather uneasy about you. When you wrote from Hampstead, you spoke uncomfortably about your health & as you have had at various times so many illnesses I should very much like to hear how you are—

If you have strength & inclination, but not without, tell me what you can about the great mapgie marriage that I may quote your account. What I call sexual selection as applied to birds has turned out to be an everlasting subject, & I am still at work on it.

Some time ago I broke down & we all went for 5 weeks to Freshwater Bay & enjoyed it much. I know you like hearing about our children so I may tell you that George has delighted us by getting his fellowship at Trinity this first year.

Poor old Sedgwick now 84 yrs old, with a multitude of complaints, most kindly wrote to congratulate us.

Leonard succeeded about half a year ago in getting in as N[o] 2. into Woolwich & means to be a R. Engineer if he can. Henrietta has been touring in Switzerland & met there two of Sir F. Darwin's daughters & much liked their frank & cordial manners.

I have nothing more to say—for this note is written to extract an account, & I much hope a better account, of yourself—

My dear Fox | your sincere friend | Ch. Darwin

P.S. I have vague remembrance that you once mentioned to me instances of Horses, or dogs, or cattle, or pigs, in which the male preferred some particular female, or conversely the female allowed of the advances of some particular male.

Have you ever kept an account, or know anyone who has done so, of number of males & females born, in pigs, dogs cattle, sheep, or *poultry*? I much want a large collection of facts on this head.

To Ernst Haeckel 19 November 1868

Down. | Bromley. | Kent. S.E.
Nov 19. 1868

My dear Haeckel

I must write to you again for two reasons. Firstly to thank you for your letter about your baby, which has quite charmed both me & my wife. I heartily congratulate you on its birth. I remember being surprized in my own case how soon the paternal instincts became developed, & in you they seem to be unusually strong. I know well the look of a baby's "hind legs" but I sh^{d} think you were the first father who had ever triumphed in their retaining a resemblance to those of a monkey What does Mrs Haeckel say to such dreadful doctrines?

I hope the large blue eyes & the principles of inheritance will make your child as good a naturalist as you are; but judging from my own experience, you will be astonished to find how the whole mental disposition of your children changes with advancing years. A young child & the same when nearly grown sometimes differ almost as much as do a caterpillar & butterfly.

The second point is to congratulate you on the projected translation of your great work—about which I heard from Huxley last Sunday. I am heartily glad of it; but how it has been brought about I know not, for a friend who supported the proposed translation at Norwich told me he thought there w^{d} be no chance of it.[1] Huxley tells me that you consent to omit & shorten some parts, & I am confident that this is very wise. As I know your object is to instruct the public, you will assuredly thus get many more readers in England. Indeed I believe that almost every book w^{d} be improved by condensation.

I have been reading a good deal of y^{r} last book, & the style is beautifully clear & easy to me; but why it sh^{d} differ so much in this respect from your great work I cannot imagine.[2] I have not yet read the first part but began with the chapter on Lyell & myself, which you will easily believe pleased me *very much*. I think Lyell, who was apparently much pleased by your sending him a copy, is also much gratified by this chapter. Your chapters on the affinities & genealogy of the animal kingdom strike me as admirable & full of original thought. Your boldness however sometimes makes me tremble, but as Huxley remarks some one must be bold enough to make a beginning in drawing up tables of descent.

Although you fully admit the imperfection of the Geological record, yet Huxley agreed with me in thinking that you are sometimes rather rash in venturing to say at what periods the several groups first appeared. I have this advantage over you that I remember how wonderfully different any statement on this subject made 20 years ago w^{d} have been to what w^{d} now be the case; and I expect the next 20 years will make quite as great a difference. . . .

I repeat how glad I am at the prospect of the translation, for I fully believe that this work & all your works will have a great influence in the advancement of Science.

Believe me my dear Häckel | your sincere friend | Charles Darwin

[1] Ernst Haeckel's *Generelle Morphologie* (Berlin, 1866) was not translated into English.

[2] Ernst Haeckel, *Natürliche Schöpfungsgeschichte* (Natural history of creation; Berlin, 1868).

To G. H. Darwin [9 December 1868]

Down.
Wednesday night—

My dear George

Many thanks for your note. I am heartily glad that dear old Backy is going on so well.[1] It gives me an awesome shudder whenever I think of it, & that is a deal too often. Tell him with my love to be sure to obey **strictly** the surgeon's directions; for I remember Engleheart said when Mrs Evans cut quite a small artery, that there was liability for longish time for the cut to open.— I am glad to hear that he is able to dissect. Tell him that I quoted with wonderful success the other day at the Nortons his proverb that "a fib in time saves nine".

I suppose you will not bring home Thompson's big Book;[2] so will you look & see exactly what he says (if he says anything) how many millions of years ago the crust of the Earth first became solidified so that it c^d^ have supported living beings. Croll quotes a passage rather too briefly for me.[3]

See if Thompson refers to any other papers by himself on subject.— I wish I knew what Haughton had said.— It is partly for this that I want Lyell's Principles,[4] & you had better bring Lyell's other book.— Croll has most kindly sent me excellent M.S. abstract of his views & a volume with *all* his own papers, which I must return, but have kept for you to read; however, I fear you will not have time on account of Wales. The brevity of the world troubles me, on account of the pre-silurian creatures which *must* have lived in numbers during endless ages, else my views w^d^ be wrong, which is impossible— Q.E.D.—

I have got Owen's book, but had not noticed the pile of abuse against me & which I must soon read. . . .[5] I daresay I shall want much advice about Croll & Thompson & be hanged to them.—

Your most affect^ly^ | C. Darwin

It is all for new Edit. of Origin.—

1 Francis Darwin had sustained a bad cut on his thigh.

2 William Thomson and P. G. Tait, *Treatise on natural philosophy* (Oxford, 1867).

3 James Croll, 'On geological time, and the probable date of the glacial and the upper miocene period', *Philosophical Magazine* 4th ser. 35 (1868): 363–84; 36: 141–54, 362–86.

4 Charles Lyell, *Principles of geology*, 10th edition (London, 1867–8).

5 Richard Owen, *On the anatomy of vertebrates* (London, 1866–8).

1869

[Carl Wilhelm von Nägeli's monograph *Entstehung und Begriff der naturhistorischen Art* (The origin and concept of natural historical species; Munich, 1865) was about the mechanisms and principles operative in the development and transmutation of species. Nägeli had augmented what he termed the theory of 'usefulness' ('Nützlichkeitstheorie', i.e., CD's theory) with his own theory of perfectibility ('Vervollkommnung'), which posited an inner tendency towards a more complex organisation or a progressive principle.]

To J. D. Hooker 13 January 1869

Down. | *Bromley.* | *Kent.*
Jan 13. 69

My dear Hooker,

As I believe you will not much grudge the trouble I send by this post 13 pages of M.S, **well written**, in answer to Nägeli for the new Edit. of the Origin. I should be extremely obliged if you would read it over, and see whether I have made any blunders, as is very likely to be the case. I have quoted Aug. St Hilaire from my own notes, & when I made them I certainly understood what he meant, but now I am not so sure that I do, and therefore am the more likely to blunder. I quote Asa Gray on Mimulus and Masters on Saponaria from the Abstracts given in the Gardiner's Chron: of papers read to the Linn. Soc. 1856 Ap. 15 & Nov. 18; and these papers I have not got; but the Abstract in the G. Chron. seems clear.

Besides looking out for blunders I should of course be grateful for any criticism, but I do not want to be unreasonable, knowing well how busy you are.

Ever yours | Ch. Darwin

The style of my M.S. will no doubt require improvement.—

[William Winwood Reade first wrote to CD in 1868, when he offered to answer queries for CD while travelling in Africa. CD sent him his Queries about expression, and also asked him particular questions about sexual selection and sexual characteristics in humans and animals.]

From William Winwood Reade 17 January 1869

C[re]. Hon Charles Heddle. | Sierra Leone

Jan[y] 17. 69.

My dear Sir

I have postponed writing to you before in the hope that I might be able to send you some information of value before I started for the interior, which I do in two or three days to search for the source of the Niger. I have unfortunately very little to say— As regards expression I can at present answer but one query. The head is shaken laterally in negative & nodded in affirmative among the Gold Coast tribes. I find it exceedingly difficult to seize expressions if not prepared for them (as one w[d]. be in case of a surgical operation). Expression is sudden & momentary— In smiling by the bye I observed two crescent wrinkles at the corners of the mouth, concave side nearest mouth— However I shall try to get something more in that way for you—

There is one breed of sheep only on the Gold Coast (where I have spent most of my time since coming out). The rams have horns & manes: the females neither. When castrated not only horns but hair is affected. I have frequently asked about the age at wh. the horns sprout but c[d]. get no answer. Europeans out here do not observe such things, & the blacks have no idea of time. However I have set three persons to watch, & hope to get a reply on my return from the interior, wh. will be in a few months. There is a breed with both sexes horned in Senegambia (with dewlaps) and another breed of enormous size in the Niger Country in wh. the female has rudimentary horns.

Women in Africa at all events among the more intelligent pagan tribes have no difficulty about getting the husbands they want, although it is considered unwomanly to *ask* a man to marry them, & that on the Gold Coast I am informed they never do— They are quite capable of falling in love, and even of forming tender passionate and faithful attachments. As a rule tribes do not intermarry but I have little doubt that there are many exceptions. Women do not like to marry strangers visiting their country because they do not like to leave their own families; on the other hand when a Foula or Mandingo comes among the Soosovo or Timnianies (pagan coast tribes in this neighbourhood) they try to make him settle or marry amongst them. They like to breed from superior men that the offspring may be of use to the town a *fact* rather opposed to the vulgar ideas about African want of foresight— Girls are not forced to marry the choice of the family in certain tribes: in others (of a lower kind) she is The intermarriage-between-tribes question is an important one. I shall pay attention to it.

I am not likely to see the gorilla up here. At present it is only found a little above Cape St. John and Loango. But so little is known about the interior that the question of its habitat, must be left open. Chimpanzees are found in the bush here. If I see one I will note down whether there is more hair on back or front.

My plans on returning from the Kong Mountains (if I get there) are not settled. I want much to study the Upper Niger & Haussa Country before I leave Africa but cannot tell whether I shall have the opportunity. I am not likely however to return home just yet. Any queries you may send to me shall be faithfully attended to, as far as lies in my power. The worst of it out here is that one gets no assistance; things which every resident ought to know one has to find out for oneself often in a few days—if one can.

Trusting that your health is permitting you to continue labouring in science, and so benefiting all who inquire and observe | I remain | Yours truly | Winwood Reade

From A. R. Wallace 20 January 1869

9, St. Mark's Crescent | NW.
Jany. 20th. 1869

Dear Darwin

It will give me very great pleasure if you will allow me to dedicate my little book of Malayan Travels to you, although it will be far too small and unpretending a work to be worthy of that honour.[1] Still, I have done what I can to make it a vehicle for communicating a taste for the higher branches of Natural History, and I know that you will judge it only too favourably.

We are in the middle of the 2nd. Vol. and if the printers will get on, shall be out next month.

Have you seen in the last Number of the "Quarterly Journal of Science", the excellent remarks on Fraser's article on Nat. Selection failing as to Man? In one page it gets to the heart of the question & I have written to the Editor to ask who the Author is.[2]

My friend Spruce's paper on Palms is to be read tomorrow evening at the "Linnæan".[3] He tells me it contains a discovery which he calls "alternation of function." He found a clump of *Geonema* all of which *were females*, and the next year the *same clump* were *all males*! He has found other facts analogous to this, & I have no doubt the subject is one that will interest you.

Hoping you are pretty well, and are getting on steadily with your next volumes,—and with kind regards to Mrs. Darwin and all your circle,

Believe me | Dear Darwin | Yours very faithfully | Alfred R. Wallace

P.S. Have you seen the admirable article in "The Guardian"! on Lyell's *Principles*? It is most excellent & liberal. It is written by Rev[d]. Geo. Buckle, of Tiverton Vicarage, Bath, who I met at Norwich and found a thoroughly scientific & liberal parson.[4] Perhaps you have heard that I have undertaken to write an article for the Quarterly! on the same subject, to make up for that on "Modern Geology" last year not mentioning Sir C. Lyell.[5] Really what with the Tories passing radical reform bills & the Church periodicals advocating Darwinianism, the Millenium must be at hand.

A R W.

[1] A. R. Wallace, *The Malay Archipelago: the land of the orang-utan, and the bird of paradise. A narrative of travel, with studies of man and nature* (London, 1869).

[2] 'The alleged failure of natural selection in the case of man', *Quarterly Journal of Science* 6 (1869): 152–3. The author, criticising an article in *Fraser's Magazine*, suggested that in human beings, as in other social animals, competition was not between individuals but between groups.

[3] Richard Spruce, '*Palmae Amazonicae*, sive enumeratio palmarum in itinere suo per regiones Americae aequatoriales lectarum', *Journal of the Linnean Society* (*Botany*) 11 (1871): 65–183.

[4] George Buckle's anonymous review of the tenth edition of Charles Lyell's *Principles of geology* (London, 1867–8) appeared in the *Guardian*, 30 December 1868.

[5] *Quarterly Review* 125 (1868): 188–217, and 126 (1869): 359–94.

To A. R. Wallace 22 January [1869]

Down. | Bromley. | Kent. S.E.
Jan 22nd.

My dear Wallace

Your intended dedication pleases me much & I look at it as a *great* honour & this is nothing more than the truth. I am glad to hear for Lyell's sake & on general grounds that you are going to write in the Quarterly. Some little time ago I was actually wishing that you wrote in the Quarterly, as I knew that you occasionally contributed to periodicals, & I thought that your articles would thus be more widely read.

Thank you for telling me about the Guardian which I will borrow from Lyell. I did note the article in the Q. Journal of Science & put it aside to read again with the articles in Frazer & the Spectator.

I have been interrupted in my regular work in preparing a new edit of the Origin, which has cost me much labour & which I hope I have considerably improved in two or three important points. I always thought individual differences more important than single variations, but now I have come to the conclusion that they are of paramount importance, & in this I believe I agree with you. Fleming Jenkyn's arguments have convinced me.[1]

I heartily congratulate you on your new book being so nearly finished be

Believe me, | My dear Wallace | yours very sincerely | Ch. Darwin

[1] [Henry Charles Fleeming Jenkin], 'The origin of species', *North British Review* 46 (1867): 277–318. See also the letter to Charles Kingsley, 10 June [1867].

From A. R. Wallace 30 January 1869

9, St. Mark's Crescent | N.W.
Jany. 30th. 1869

Dear Darwin

Will you tell me *where* are Fleming Jenkyns' arguments on importance of single variations. Because I at present hold most strongly the contrary opinion, that it is the individual differences or *general variability* of species that enables them to become modified and adapted to new conditions.

Variations or "sports" may be important in modifying an animal in one direction, as in *colour* for instance, but how it can possibly work in changes requiring co-ordination of many parts, as in *Orchids* for example, I cannot conceive. And as all the more important structural modifications of animals & plants imply much coordination, it appears to me that the chances are millions to one against *individual variations* ever coinciding so as to render the required modification possible.

However let me read first what has convinced you.

You may tell Mrs. Darwin that I have now a daughter.

Give my kind regards to her & all your family.

Very truly yours | Alfred R. Wallace—

[William Thomson, the leading physicist of the time, had charged that the great length of time needed for the evolution of species according to CD's theory was in direct conflict with the age of the earth as established by the laws of physics as they were then understood. No answer to this charge was possible until the discovery of radioactivity and the establishment of the timescale provided by atomic decay.]

To James Croll 31 January [1869]

Down. | Bromley. | Kent. S.E.
Jany 31

My dear Sir

Tomorrow I will return registered your book, which I have kept so long.

I am most sincerely obliged for its loan, & especially for the MS. without which I sh^{d} have been afraid of making mistakes.[1]

If you require it the M.S shall be returned. Your results have been of more use to me than I think any other set of papers which I can remember.

Sir C. Lyell who is staying here is very unwilling to admit the greater warmth of the S. hemisphere during the glacial period in the N; but, as I have told him, this conclusion, which you have arrived at from physical considerations, explains so well whole classes of facts in distribution, that I must joyfully accept it; indeed I go so far as to think that your conclusion is strengthened by the facts in distribution. Your discussion on the flowing of the great ice-cap southward is most interesting.

I suppose that you have read Mr Moseley's recent discussion on the force of gravity being quite insufficient to account for the downward movement of glaciers: if he is right, do you not think that the unknown force may make more intelligible the extension of the great northern ice cap.[2] Notwithstanding your excellent remarks on the work which can be effected within a million years, I am greatly troubled at the short duration of the world according to Sir W. Thompson, for I require for my theoretical views a very long period *before* the Cambrian formation. If it w[d] not trouble you I sh[d] like to hear what you think of Lyell's remarks on the magnetic force which comes from the sun to the earth; might not this penetrate the crust of the earth & then be converted into heat. This w[d] give a somewhat longer time during which the crust might have been solid; & this is the argument on which Sir W. Thompson seems chiefly to rest. You seem to argue chiefly on the expenditure of energy of all kinds by the sun, & in this respect Lyell's remark wd have no bearing.

My new edition of the "origin" will be published, I suppose in about two months, & for the chance of y[r] liking to have a copy, I will send one.

With my very sincere thanks for all your kind assistance | I remain | yours very faithfully | Charles Darwin

I wish that you would turn your astronomical knowledge to the consideration whether the form of the globe does not become periodically slightly changed, so as to account for the many repeated ups & downs of the surface in all parts of the world.— I have always thought that some cosmical cause would some day be discovered

1 Croll had sent CD a manuscript summarising his views on ice ages and global temperature, together with a book containing all of his published papers up to that time and possibly some unpublished work.

2 Henry Moseley, 'On the mechanical possibility of the descent of glaciers by their weight only', *Scientific Opinion* 1 (1869): 191–2.

From James Croll 4 February 1869

Edinburgh
Feby 4th 1869.

Dear Sir,

Your favour with book came duly to hand, and I am glad to hear that some of the papers have been of a little use to you. I am very much pleased to learn that you consider the facts in distribution favourable to some of the views expressed in my papers on climate

I have not as yet been able to overtake that part of the question relating to the condition of the hemisphere whose winters occur in perihelion. I have no doubts that when this part of the subject has been fully discussed Sir Charles Lyell will agree with me. The facts in favour of a warm climate are so numerous and strong

It is a pity that Sir Charles should have made those remarks on the "secular loss of heat in the solar system" vol. II p 213.[1] He must have done it without due consideration of that point

If there is one thing more than another in physics, regarding which we have absolute certainty, it is that the solar system is losing its store of energy. We not only know this fact, but we have a means of determining the actual rate at which it is losing its power. 3,869,000 foot-pounds of energy in the form of heat is radiated off every square foot of the sun's surface per second. In other words the quantity of energy thrown off into space by the sun is equal to a *7000 horse* power engine working on every square foot of its surface. And when we reflect that all this prodigious expenditure has been going on during countless geological ages we may well ask the question what is the secret of the sun's great strength? Gravitation only affords up to the present time 20,000,000 years heat There must be some other source in addition to that of gravitation. It is strange that that other possible sources did not suggest itself to Sir William Thomson and other physians [physicists] when working at this question. It is perfectly obvious that the sun or rather the matter which composes the sun might have been in possession of heat prior to condensation In this case it is difficult to say how old the sun may be for we do not know what this original store may have amounted to. In my paper I assumed a certain relation between the amount of original heat and that produced by gravitation viz 234 to 95 (Phil. Mag May 1868) but as I stated, I may be wrong It may be more than this, or it may be less. This proportion gives 70 000 000 years

The introduction of this new element, changes the entirely the conditions of the problem and I have no doubt that the whole matter will have to be re-considered. And it is quite possible that we may yet be able to get considerably more than one hundred millions of years although very much beyond this we are brought to a limit by other considerations.

As regards determining the age of the earth's crust from the secular cooling of the globe I am not altogether satisfied with the plan. It would no doubt do if we had proper data to go by, but I don't think we have got that yet.

I think that you may quite fairly assume a very long period before the Cambrian formation, even according to Sir William Thomson's theory for supposing the earth to have orginally been in a molten condition, a solid crust would very rapidly form, and if this crust would not break up and sink, the globe at the surface would be cool and suitable for life although a short way down below the surface the heat was intense. This results from the slow rate at which the crust is able to conduct the heat from within

It is some years since I read Sir William's paper on the secular cooling of the globe, but I think he states the above as his opinion, at all events I heard him once say in a lecture on the subject that supposing the earth to be in a molten state, in a few thousand years you could walk on its surface and hardly be sensible of the heat from within[2]

Electricity and Magnetism used to be my favourite study,, but for the past four years I have been paying little attention to what was going on in that department.

A relation between the spots of the sun and the manifestation of electric phenomena on the earth does not *necessarily imply any transmission of* electric forces from the one body to the other

One thing is certain that it is but an infinitesimal quantity of the forces of nature that ever assume the electric or magnetic form. Electrical phenomena is very imposing and this is the reason why so much is attributed to it. A thunder storm is something very striking but Faraday has shown that more electricity is evolved in the silent decomposition of a few *grains* of water in the cell of a battery than would be required to produce the most violent flash of lightning

It is owing to *high tension* that electricity makes such a display in passing from the statical to the dynamical state But when you estimate the amount of energy thus displayed in foot-pounds it is often very little

The quantity of energy in the form of electricity coming from the sun (if there be any at all) is certainly trifling compared with what comes in the form of heat. I believe that no physicist will call this in question.

Your suggestion as to the possibility of a cosmical cause for the ups and downs of the crust never occurred to my mind. I can see no possible way at present how the thing can be, but I shall ponder certainly over it.

I have never heard of Mr Moseley's paper. My curiosity is very much excited and perhaps you will be so kind as to let me know when the paper appeared.

Edinburgh with all its books and learning is miserably behind in scientific literature. Since I came here I hardly know what is going on in the scientific world around. One can get plenty of good solid books on science, but the current news and literature of the subject is not to be found anywhere. Edinburgh I fear is falling behind.

I need hardly say that a present of a copy th[e] Origin of Species from its Author will be esteemed worth a dozen of copies out of a shop.

I trust you will make out to read this rather long affair, written hurriedly to catch the post.

I am, | yours very truly, | James Croll.

Charles Darwin— Esqr. FRS—

Keep the M.S sent it is of no use to me.

[1] Charles Lyell, *Principles of geology*, 10th edition (London, 1867–8).

[2] In 'On the secular cooling of the earth' (*Tranactions of the Royal Society of Edinburgh* 23 (1864): 157–70), William Thomson set the period of the consolidation of the earth at more than twenty million but less than four hundred million years. CD did not think this would be long enough to allow for the origin of species according to his theory.

To Thomas Roscoe Rede Stebbing 3 March 1869

Down. | Bromley. | Kent.
March. 3rd. 1869

Dear Sir,

I am very much obliged to you for your kindness in sending me your spirited & interesting lecture.[1] If a layman had delivered the same address, he would have done good service in spreading what, as I hope and believe, is to a large extent the truth; but a clergyman in delivering such an address does, as it appears to me, much more good by his power to shake ignorant prejudices, & by setting, if I may be permitted to say so, an admirable example of liberality.

With sincere respect I beg leave to remain— | Dear Sir, | Yours faithfully & obliged | Charles Darwin

[1] T. R. R. Stebbing, *Darwinism; a lecture delivered before the Torquay Natural History Society* (London, 1869).

From T. R. R. Stebbing 5 March 1869

Tor Crest Hall, | Torquay.
March 5th. 1869.

My dear Sir,

It may be interesting to you to know that I began the perusal of your works with the common prejudices strong upon me, and a very impertinent expectation that I should find a damaging flaw somewhere or another in your argument.

It was a vast surprize to me and soon became a great pleasure to find what was the real character of these insidious writings, which even men of considerable scientific attainments seemed anxious to taboo. Of course I had to undergo the useful pang of giving up some opinions on which I had formerly been very positive; but in return there was laid before me a view of the world's history, so simple, so harmonious, so unlike the chaotic shreds and patches that most histories are made up of, that I should have felt genuine admiration, even had I not been also convinced.

Incidentally, I may add, it was no slight pleasure to "an orthodox clergyman" to find in his Author's calm and temperate style that Science might make gigantic strides without offering such collateral opinions as, if true, would certainly dispense with clergymen altogether, whether orthodox or otherwise.

The kind letter in which you approve my slight attempt to get people to study your works for themselves has afforded me one more gratification.

Believe me | Yours most faithfully | Thomas R. R. Stebbing . . .

[In 1869, an English translation was published of Fritz Müller's *Für Darwin*, a study of the developmental history of the Crustacea that Müller presented as a validation of CD's theory of natural selection. The translation (*Facts and arguments for Darwin*) was made at CD's behest by William Sweetland Dallas, the indexer of *Variation*.]

To Fritz Müller 14 March 1869

Down. | *Bromley.* | *Kent. S.E.*
March 14th 1869/

My dear Sir

I fear that your patience will have been completely exhausted; but at last the Translation is published. The delay has been caused by the Translator; but he is not to blame for he had to move a large family several hundred miles & entered on his new office at the busiest time of the year, so that it was really impossible for him to do anything. I have this day sent off 3 copies to you by Book Post.

I do hope that you will be contented with the appearance & with the Translation. I have not read it yet, but Dallas generally translates well & was greatly interested with your work.— I have sent copies "From Author" to C. Spence Bate, Dana in U. States, your Brother H. Müller, Max Schu[l]tz of Bonn & Oscar Schmidt at Gratz.— I am ashamed to say that I cannot find your instructions about the presentation copies, though they are somewhere safe, so I wrote to your brother & he suggested the two last names.— If you will write I will send copies to anyone else & of course more to yourself. I have sent copies to 7 of such reviews, as generally treat of Scientific works.— My publisher can form no idea whether the book will sell, but fears it is too purely scientific for England. There will be 1000 copies printed.

I received some time ago a very interesting letter from you with many facts about Oxalis, & about the non-seeding & spreading of one species.— I may mention that our common *O. acetosella* varies much in length of pistils & stamens, so that I at first though[t] it was certainly dimorphic, but proved it by experiment not to be so.— Borreria has after all seeded well with me when crossed by opposite form, but very sparingly when self-fertilised. Your case of Faramea astonishes me.— Are you sure there is no mistake— the difference in size of flower & wonderful difference in size & structure of pollen-grains naturally makes one rather sceptical.

I never fail to admire & to be surprised at the numbers of points to which you attend.

I go on slowly at my next book & though I never am idle, I make but slow progress, for I am often interrupted by being unwell & my subject of sexual selection has grown into a very large one.— I have, also, had to correct a new Edit. of my "Origin", & this has taken me six week, for science progresses at Rail-road speed. I cannot tell you how rejoiced I am that your Book is at last out; for whether it sells largely or not, I am certain it will produce a great effect on all capable judges, though these are few in number.

Believe me | Yours sincerely & cordially | Ch. Darwin

(I am vexed to see that on Title my name is more conspicuous than yours, which I especially objected to & cautioned Printers, after seeing one Proof.—)

I see that the seed that the nurseryman has sent me is that of Eschotzia crocea.— Perhaps my observations & yours have been made on distinct species!!!!

P.S. I have just received your letter of Jan. 12th.— I am greatly interested by what you say on Eschotzia— I wish your plant had succeeded better. It seems pretty clear that the species is much more self-sterile under the climate of Brazil than here, & this seems to me important result.— I have no spare seeds at present, but will send for some from nurseryman, which though not so good for our purpose will be worth trying. I can send some of my own in autumn You could simply cover up *separately* 2 or 3 single plants, & see if they will seed without aid,—mine did abundantly.— ...

With sincere admiration | Yours very truly | C. Darwin

P.S. I have sent one copy to Mr Wallace, as I am sure he wd enjoy the work & cannot afford to buy many Books ...

To T. H. Huxley 19 March [1869]

Down. | Bromley. | Kent. S.E.
March 19th

My dear Huxley

Thanks for your Address.[1]

People complain of the unequal distribution of wealth, but it [is] a much greater shame & injustice that any one man shd. have the power to write so many brilliant essays as you have lately done. There is no one who writes like you.— I have hugely enjoyed your attack on Thompson; but if I were in your shoes I should tremble for my life.— I agree with all you say, except that I must think that you draw too great a distinction between your evolutionists and the uniformitarians.

I find that the few sentences which I have sent to press in the Origin about the age of the world will do fairly well, though if I had read you first, perhaps I sh^{d} have been less deferential towards Thompson.[2]

Many thanks for your note received a day or two ago with yourself represented as a bristly little Terrier.—

Ever yours | C. Darwin

[1] As president of the Geological Society of London, Huxley delivered the 1869 anniversary address. In it he criticised William Thomson's argument that life could not have existed on earth for more than 'some such period of time as one hundred million years'. CD regarded Thomson's argument as a threat to his theory of the origin of species.

[2] In *Origin* 5th ed., p. 354, CD wrote, 'According to the standard of years we have no means of determining how long a period it takes to modify a species. Mr. Croll judging from the amount of heat-energy in the sun and from the date which he assigns to the last glacial epoch, estimates that only sixty million years have elapsed since the deposition of the first Cambrian formation. This appears a very short period for so many and such great mutations in the forms of life, as have certainly since occurred. It is admitted that many of the elements in the calculation are more or less doubtful, and Sir W. Thomson gives a wide margin to the possible age of the habitable world. But as we have seen, we cannot comprehend what the figures 60,000,000 really imply; and during this, or perhaps a longer roll of years, the land and the waters have everywhere teemed with living creatures, all exposed to the struggle for life and undergoing change.'

[After much encouragement from CD, Alfred Russel Wallace at last published his *Malay archipelago* (London, 1869), the full account of his travels 'in the land of the orang-utan and the bird of paradise'. Like CD's *Journal of researches*, it became one of the great natural-history travel books.]

To A. R. Wallace 22 March [1869]

Down. | Bromley. | Kent. S.E.
March 22^{d}

My dear Wallace

I have finished y^{r} book; it seems to me excellent, & at the same time most pleasant to read. That you ever returned alive is wonderful after all y^{r} risks from illness & sea voyages, especially that most interesting one to Waigiou & back. Of all the impressions which I have rec^{d} from y^{r} book the strongest is that y^{r} perseverance in the cause of science was heroic. Your descriptions of catching the splendid butterflies have made me quite envious, & at the same time have made me feel almost young again, so vividly have they brought before my mind old days when I collected, tho' I never made such captures as yours. Certainly collecting is the best sport in the world. I shall be astonished if y^{r} book has not a great success; & your splendid generalizations on Geog. Distrib., with which I am familiar from y^{r} papers, will be new to most of your readers. . . .

Many parts of y^{r} book have interested me much: I always wished to hear an independent judgment about the Rajah Brooke, & now I have been delighted with y^{r} splendid eulogium on him.

With respect to the fewness & inconspicuousness of the flowers in the Tropics, may it not be accounted for by the hosts of insects, so that there is no need for the flowers to be conspicuous. As according to Humboldt fewer plants are social in the tropical than in the temperate regions, the flowers in the former w^{d} not make so great a show.

In y^{r} note you speak of observing some inelegancies of style, I notice none. All is as clear as daylight. . . .

I have only one criticism of a general nature, & I am not sure that other geologists w^{d} agree with me: you repeatedly speak as if the pouring out of lava &c from Volcanoes actually caused the subsidence of an adjoining area; I quite agree that areas undergoing opposite movements are some how connected; but volcanic outbursts must I think be looked at as mere accidents in the swelling up of a great dome or surface of *plutonic* rocks,—& there seems no more reason to conclude that such swelling or elevation in mass is the cause of the subsidence, than that the subsidence is the cause of the elevation; which latter view is indeed held by some geologists.

I have regretted to find so little about the habits of the many animals which you have seen.

In Vol 2. p. 399 I wish I c^{d} see the connection between variations having been first or long ago selected & their appearance at an earlier age in birds of Paradise than the variations which have subsequently arisen & been selected. In fact I do not understand y[r] explanation of the curious order of development of the ornaments of these birds.

Will you please to tell me whether you are sure that the female Casuarius (Vol. 2. 150 sits on her eggs as well as the male; for if I am not mistaken Bartlett told me that the male alone, who is less brightly coloured about the neck, sits on the eggs.

In Vol. 2 p. 255 you speak of male savages ornamenting themselves more than the women, of which I have heard before; now have you any notion whether they do this to please themselves, or to excite the admiration of their fellow men, or to please the women, or, as is perhaps probable from all 3 motives.

Finally let me congratulate you heartily on having written so excellent a book, full of thought on all sorts of subjects. Once again let me thank you for the very great honour which you have done me by your dedication

Believe me | My dear Wallace | Yours very sincerely | Ch. Darwin

Vol. 2. p. 455 When in New Zealand I thought the inhabitants a mixed race, with the types of Tahiti preponderating over some darker race with more frizzled hair; & now that the stone instruments reveal the existence of ancient inhabitants is it not probable that these Is[d] were inhabited by true Papuans. Judging from descriptions the pure Tahitians must differ much from your Papuans.—

From Harrison William Weir 23 March 1869

Weirleigh | Brenchley Kent.
March 23rd. 1869

Dear Sir

Since I last wrote to you I have been collecting information as to the numbers of the different sexes in the families of birds and Animals

As regards Pigs *Most* of the breeders that I have consulted on the matter tell me that the average of males to females is about 7 to 6, but my own observation though limited makes them nearly if not quite 8 males to 6 females and one litter that I met with of **13** (of which I purchased 2) there were no less than **12** males and only **1** female. Once at Halifax I was shewn one of *neither* sex a little of *each*.

As regards Pigeons

There are more cocks than hens, I have often bred two cocks in a nest, but seldom two hens. The hen in the nest is generally the weaker bird of the two, and is much more liable to die in rearing. Supposing that you were breeding from two blues, the hen of the two young ones frequently (perhaps *not* very so) comes out a *silver* color, but I do not remember a visa versa case. In breeding from two reds the hen in nest is sometimes *yellow* while the cock bird is red.

With regards to the likes and dislikes

Some of the fanciers have informed me that their hens will often take a fancy to one particular cock and often leave their own mates this I have also known to be the case with my own birds but not frequent

Pheasants.

M^r. Baker of Leadenhall market tells me that in breeding the above, he generally gets *4* to *5* Cocks to 1 hen. When out shooting (where we have killed *all rising* or rather shot at) *I* have noticed the great preponderance of cocks, when the bag has been laid out.

Pigeons

M^r. Ridpeth of Manchester keeping blue Rock pigeons has noticed that they *drive off* any *other color*, such as yellow, reds, and whites

Tame Rabbits.

I have kept rabbits many years, and have noticed the far greater number of bucks than does. I kept an account last year 1868 on purpose for you and out of **three** litters **15** in all I only had *one* doe. I will keep an acc^t again *this* year.

I have had my Rabbits some generations back but curiously enough I bred a **long** hair last year It is **grey** and a buck. The hair is *very* long. I have kept it on purpose to see if the breed from it will be long haired. I have one litter now about 6 weeks old from it *all* smooth & shall try again as an experiment or if you would like to have it I shall be most happy to send it you. I found that the doe rabbits did not care so much for his attentions as the smooth buck *Rabbits* are different from any other any animals I know. They will receive the attention of the buck *though with young*, even up to the *day before* kindling and if put to different bucks seem not particular as to how many this is curious so *excuse* my mentioning it as I know of no other animal like it.

Breeding singing birds

I am informed that the best *singing* bird (cock) generally gets a mate first when they are bred in rooms.

A Poultry breeder told me a few days since (He keeps Bramah Poutras and Spanish) that *some years* his Spanish *run mostly* to cocks and the Bramahs the opposite. Then the next year the Bramah are mostly cocks possibly, but he seems to have the idea that there is something in the *season* that helps the breeding in some way, this (if true) seems to carry out my notice of the sheep some seasons having more twins

Incomplete

To A. R. Wallace 14 April 1869

Down. | *Bromley.* | *Kent. S.E.*
Ap. 14. 1869

My dear Wallace

I have been wonderfully interested by your article, & I sh^{d} think Lyell will be much gratified by it.[1] I declare if I had been editor & had the power of directing you I sh^{d} have selected for discussion the very points which you have chosen. I have often said to younger geologists (for I began in the year 1830) that they did not know what a revolution Lyell had effected; nevertheless y^{r} extracts from Cuvier have quite astonished me. Though not able really to judge, I am inclined to put more confidence in Croll than you seem to do; but I have been much struck by many of y^{r} remarks on degradation. Thompson's views of the recent age of the world have been for some time one of my sorest troubles, & so I have been glad to read what you say. Your exposition of Nat. selection seems to me inimitably good; there never lived a better expounder than you. I was also much pleased at y^{r} discussing the difference between our views & Lamarck's. One sometimes sees the odious expression "Justice to myself compels me to say &c"; but you are the only man I ever heard of who persistently does himself an injustice & never demands justice. Indeed you ought in the review to have alluded to y^{r} papers in Linn. Journal,[2] & I feel sure all our friends will agree in this. But you cannot "Burke" yourself, however much you may try, as may be seen in half the articles which appear. I was asked but the other day by a German Prof. for y^{r}

paper which I sent him. Altogether I look at y^r article as appearing in the Q—ly as an immense triumph for our cause. I presume that y^r remarks on Man are those to which you alluded in y^r note.

If you had not told me I sh^d have thought that they had been added by some one else. As you expected I differ grievously from you, & I am very sorry for it. I can see no necessity for calling in an additional & proximate cause in regard to Man. But the subject is too long for a letter. I have been particularly glad to read y^r discussion because I am now writing & thinking much about man.

I hope that y^r Malay book sells well: I was extremely pleased with the Art. in the Q. J. of science,. inasmuch as it is thoroughly appreciative of y^r work: Alas! you will probably agree with what the writer says about *the uses of the bamboo.*[3]

I hear that there is also a good article in the Sat. Rev., but have heard nothing more about it.

Believe me my dear Wallace | yours ever sincerely | Ch. Darwin

P.S. I have had a baddish fall. my horse partly rolling over me, but I am getting rapidly well—

1 [A. R. Wallace], 'Sir Charles Lyell on geological climates and the origin of species' (review of Lyell's *Principles of geology*, 10th ed. (1867–68), and *Elements of geology*, 6th ed. (1865)), *Quarterly Review* 126 (1869): 359–94.

2 CD refers principally to his and Wallace's papers, which were jointly published under the title, 'On the tendency of species to form varieties; and on the perpetuation of varieties and species by natural means of selection', *Journal of the Proceedings of the Linnean Society* (*Zoology*) 3 (1859): 45–62. On the history of this paper, see *Charles Darwin's letters: a selection 1825–59* (Cambridge, 1996), pp. xix–xx, 188–98. Wallace had published other papers in the *Proceedings* and the *Transactions* of the Linnean Society of London.

3 *Quarterly Journal of Science* 6 (1869): 172: 'if we wish for a mass of evidence in favour of design, before which Paley pales, we need only read the author's account of the Bamboo and its uses. . . . He shows that it is indispensable to the natives. Looking at their mental condition, they could not have existed without it, or some similar boon of Providence.'

[Julius Victor Carus succeeded Heinrich Georg Bronn as German translator of *Origin* when he revised Bronn's translation for the third German edition in 1867. He translated subsequent editions of *Origin*, and CD's other works, into German; his translation of the fifth English edition of *Origin* was published in 1872.]

To Julius Victor Carus 4 May 1869

Down. | Beckenham | *Kent. S.E.*
May 4. 1869

My dear Sir

I am pleased to hear about the Origin, for I am now completing a new edit., which will be published in a month's time or less.[1] I have gone very carefully thro' the whole, trying to make some parts clearer, & adding a few discussions & facts of some importance. The new ed. is only 2 pages, at the end, longer than the old; though in one part 9 pages in advance, for I have condensed several parts, & omitted some passages. The translation, I fear will cause you a good deal of trouble; the alterations took me 6 weeks, besides correcting the press; you ought to make a special agreement with M. Koch. Many of the corrections are only a few words, but they have been made from the evidence on various points appearing to have become a little stronger or weaker.

Thus I have been led to place somewhat more value on the direct & definite action of external conditions,—to think the lapse of time as measured by years not quite so great as most geologists have thought,—& to infer that single variations are of even less importance, in comparison with individual differences, than I formerly thought. I mention these points because I have been thus led to alter in many places *a few words*; & unless you go thro' the whole new edition, one part will not agree with another, which w^{d} be a great blemish. I c^{d} lend you the corrected sheets, which w^{d} shew where changes had been made, but then some corrections have been added in the proofs, which w^{d} not appear in the above sheets. Please sometime to inform me whether you wish for the loan of the corrected sheets, & whether you w^{d} like a *bound* copy, or *clean loose* sheets for y^{r} translation.

I am pleased to hear about the sale of "Domestic Animals."[2] Altho' I am never idle my new book makes very slow progress; & I will assuredly inform you as soon as it goes to the printers.

Believe me | my dear Sir | yours very sincerely | Charles Darwin

[1] Carus had written to CD that the publishers of the German translation of *Origin* had informed him that a new edition was required.

[2] *Variation.*

To J. D. Hooker 22 June [1869]

Caerdeon, Barmouth | N. Wales
June 22^{d}.

My dear Hooker

It was very good of you to write to me from Stockholm, telling me all the things about which I liked to hear.—[1] I had heard nothing of your doings, except, as stated in Gard.Chronicle, that the announcement of your being President of the next Congress, was received with loud & general applause; & this applause made me applaud the meeting.—

I suppose the Emperor could have given you nothing which you would have liked better than the Vases, which even I sh^{d}. like to see: he must have heard of your Crockery madness!—[2]

We have been here for 10 days: how I wish it were possible for you to pay us a visit here: we have a beautiful House, with a terraced garden, & a really magnificent view of Cader, right opposite. Old Cader is a grand fellow & shows himself off superbly with every changing light. We remain here till end of July, when the H. Wedgwoods have the house.— I have been as yet in a very poor way: it seems as soon as the stimulus of mental work stops, my whole strength gives way: as yet I have hardly crawled half a mile from the House, & then been fearfully fatigued.— It is enough to make one wish oneself quiet in a comfortable tomb.

I suppose that you have read Bentham's Address:[3] it has interested me greatly, as I particularly wished to hear how Botanists agreed with Zoologists about Distribution. Everything Bentham says, always seems to me remarkably *wise*; & I have an instinctive feeling that every hint from him is well worth pondering over.— Nevertheless I must still think more of the importance of isolation in preserving old Forms, than he seems inclined to do.— — Does not this Address make you wish to write another essay like your former splendid ones?

I am glad to hear about Andersson & Galapagos: I w^{d}. gladly subscribe, & if necessary w^{d} go as far as 50£.— He surely ought to visit Cocos Isl^{d}.— Has anyone ever examined the Revillagagos or some such name Arch. off Mexico? I remember looking in old days with longing eyes at this group on the Chart.—

I fear that you forgot all about the colour of the Beards of the Sclavonic races relatively to the colour of the Hair.— Give our very kind remembrances to M^rs. Hooker, & I am very glad to hear that she enjoyed your tour, which seems to have been in every way a most successful affair—

Yours most affect. | C. Darwin

[1] Hooker had attended an international congress in St Petersburg.

[2] Tsar Alexander II sent Hooker two jasper vases in lieu of a decoration that Hooker had declined. Hooker was an avid collector of Wedgwood chinaware.

[3] George Bentham, 'Anniversary address: on geographical biology', *Proceedings of the Linnean Society of London* 10 (1869): lxv–c.

From Albert Günther 23 September 1869

British Museum
23.9.69

My dear Sir

In order to show you how much pleasure I have in answering your queries, I express a hope that this is merely the first batch of questions

I am glad to say that Mr Ford will be able to complete your drawings within 2 months, & he has been at work at them today. With regard to the heads of Salmon, a figure of a *male S. lycaodon* can be copied from Richardson, but then we have no female of this species. Therefore I should advise you to take a British Salmonoid which will answer your purpose perfectly. Again, you require the figures of males & females of exactly the same size, but then one sex is sometimes the larger, as in *Callion. draco*, & therefore your figures would be faulty in this respect. I wait for your instructions.

It is but a short time ago that you told me that a good wife is the greatest blessing for a man. How often did I remember these words, when I hurried away from trouble & work home where peace & happiness awaited me. Alas! it was a short dream in life's long night—my wife who was the greatest blessing to me every day of our married life, has been taken away from me. Last month she gave birth to a strong & healthy child, a boy; all appeared to go well, when the excessive heat came, & with it a fever which hurried her off into an early grave.

A time & a trial like this make a man think of affairs beyond the grave; and that men must be strong who can rest satisfied with the glimpses or the semblance of truth gained by man's speculation. I confess I am weak enough to require better comfort, & this I can only find in faith into truth revealed

Yours ever faithfully | A Günther

To Albert Günther 25 September 1869

Down. | Beckenham | *Kent. S.E..*
Sept. 25./69

My dear Dr. Günther

I have just received your letter, & am astonished & deeply grieved at its contents. You have my most entire sympathy. Your words so full of tenderness & resignation brought tears into my eyes.

There is at present no comfort for you, & I can only wish you fortitude to bear the greatest loss, which a man can ever have to bear.—

I will write in few days & in the meantime I thank you for all your very kind assistance, so kindly rendered.—

Believe me, Yours very sincerely | Ch. Darwin

To John Brodie Innes 18 October 1869

Down. | Beckenham | *Kent. S.E.*
Oct. 18 | 69

My dear Innes

I was wishing to hear some news of you, & had thought of writing, but I get so many foolish letters from foolish people, that I seldom have the heart to write to my friends. There is hardly any news to tell you of your old Parish.— Mr Powell has taken Mr Engleheart's House & this I am very glad of, as he will now be able to look after the Parish & school, & I daresay he will be active & kind; but I rather doubt whether he is above the average in sense.— We suspect on very slight grounds that he is going to be married; for he has given notice to Amy Duberry that a Lady will soon teach in the Sunday School.— Possibly it may be Mrs Lovegrove.— I hear of no chance of a parsonage being built; Mr Powell wished to get up a subscription, but I doubt whether he will succeed.— I offered 20 £ which he seemed to think very small, but I shall not increase the amnt; for I see no reason that the Parsonage shd cost 16 or 1700 £, as he proposes. Mr Engleheart's lungs (& I fear purse also) failed him; & he is to us a fearful loss as a doctor.

I have neither seen nor heard anything of the Lubbocks for an age—but this is not true for I often hear their Harriers in the morning.

M^{rs}. H. Lubbock finds Gorringes so dull that they intend taking a London House, & coming occasionally to the Farm for the hunting in the winter; & I suspect poor Henry Lubbock is ready to hang himself at the thought of his London life.—

I have not seen or heard of Mr Huttons[1] remarks on me at the Liverpool congress.; & this I regret, for I suppose it is Mr Hutton, editor of the Spectator, who is a very clever man, who feels a deep interest in religion, though he w^{d} be considered by all Churchmen as highly Latitudinarian.— The newspapers have lately been abusing, praising & chaffing me at a great rate.—

1 Richard Holt Hutton.

To Francis Galton 23 December [1869]

Down. | Beckenham | *Kent. S.E.*
Dec. 23^{d}

My dear Galton

I have only read about 50 pages of your Book (to the Judges)[1] but I must exhale myself, else something will go wrong in my inside. I do not think I ever in all my life read anything more interesting & original. And how well & clearly you put every point! George, who has finished the Book, & who expressed himself just in the same terms, tells me the earlier chapters are nothing in interest to the latter ones! It will take me some time to get to those latter chapters,, as it is read aloud to me by my wife, who is also much interested.— You have made a convert of an opponent in one sense, for I have always maintained that, excepting fools, men did not differ much in intellect, only in zeal & hard work; & I still think there is an *eminently* important difference.

I congratulate you on producing what I am convinced will prove a memorable work.—

I look forward with intense interest to each reading, but it sets me thinking so much that I find it very hard work; but that is wholly the fault of my brain & not of your beautifully clear style.—

Yours most sincerely | Ch. Darwin

1 Francis Galton, *Hereditary genius: an inquiry into its laws and consequences* (London, 1869). The 'judges' were probably Emma and Henrietta Darwin.

1870

To Anton Dohrn 4 January 1870

Down. | Beckenham | *Kent. S.E.*
Jan 4 1870

My dear Sir

On the receipt of your letter I wrote at once to the Director of our Zoolog. Gardens, about the Limulus. He informs me that they now have only one old adult male; but when they had both sexes eggs were *never* observed. Your best chance w^{d} be to write to some Zoologist on the shores of the United States; I cannot help you in this, as my sole correspondent, Dana, is now much out of health.

I enclose a separate letter about your scheme, which has my good wishes; but I am sure that you estimate my influence & judgment much too highly.[1]

The opinion of naturalists who have visited the coast for some special investigation w^{d} be worth far more than mine. I fear your plan will cost you much loss of time in writing letters & making arrangements. I w^{d} suggest to you to delay attempting so great an addition as the formation of a scientific library.

Lastly let me thank you for your very kind & strong expressions towards myself, & for your information with respect to your present views on embryology.

Forgive me for suggesting one caution; as Demosthenes said, "action, action, action" was the soul of eloquence, so is caution almost the soul of science.

Pray bear in mind that if a naturalist is once considered, though unjustly, as not quite trust worthy, it takes long years before he can recover his reputation for accuracy.

Pray forgive me for this caution. You ask about my health; I cannot say much in its praise, but as long as I live the life of a hermit I am able to work some hours daily.

With the most sincere good wishes | I remain my dear Sir | yours very faithfully | Ch. Darwin

[Enclosure]

Down. | Beckenham | *Kent. S.E.*
Jan 4 1870

My dear Sir

In your letter of Dec. 30 you ask me for my opinion with respect to founding an acquarium with the necessary apparatus, at some favorable station such as Messina, for scientific researches.

As far as my judgement goes, I can feel no doubt that at present embryological investigations on the lower marine animals are of the utmost importance; & for this purpose your scheme offers obvious facilities. If Johann Müller[2] c^{d} have left the larvæ of the Echinoderms in an acquarium well provided with a flow of sea-water, he c^{d} have examined them after considerable intervals of time, or he might have arranged with some other naturalist to follow out & mature his great discovery. Therefore if sufficient funds can be obtained to construct & keep up an acquarium, & if it is found practicable to regulate its use amongst old & new subscribers or strangers, you will, no doubt have the good wishes of every naturalist in Europe. For the sake of shewing my own good will, I shall at any time be happy to subscribe the small sum of five pounds sterling towards the necessary expenses

Pray believe me my dear Sir | yours very faithfully | Charles Darwin . . .

1 Dohrn was planning to set up a zoological station in Italy. His plan met with support and the station was founded in 1873 with Dohrn as director.

2 Johannes Peter Müller.

[CD's daughter Henrietta was frequently employed by him as a critical reader of his manuscripts. This letter refers to chapters 3 and 4 of *Descent.*]

To H. E. Darwin [before 17 February 1870]

Spring 1870

My dear H.

Please read the Ch. first **right through** without a pencil in your hand, that you may judge of general scheme;, also, I particularly wish to know whether parts are extra tedious; but remember that M.S is always *much* more tedious than print.— The object of Ch. is simply co[m]parison of mind in men & animals:

in the next chapt. I discuss progress of morals &c.— Some sentences are at back of Page marked thus @.—

I do not send foot-notes, as I have no copy & they are almost wholly mere authorities.— After reading once right through, the more time you can give up for deep criticism or corrections of style, the more grateful I shall be.— Please make any long corrections on separate slips of paper, leaving narrow blank edge, & pin them to margin of each sheet, so that I can turn each back, & read whilst still attached to its proper page.— This will save me a world of trouble Heaven only knows what you will think of the whole, for I cannot conjecture.— You are a very good girl indeed to undertake the job.—

Your affect Father | C. Darwin

(I suspect that here & there style will want a good deal of improvement, though I hope greater part fair.—)

(I fear parts are too like a Sermon: who w^d ever have thought that I sh^d. turn parson?)

From H. E. Darwin [17 February 1870]

Dear F.

Thanks for your Note abt the M.S.S. I will obey all your instructions—& I shall be extremely interested in reading it—so you needn't thank me for it—& here what time I spend in my own room is so very undisturbed it goes much further. When I know no human being will come after me & lay out my plans for the day & stick to them. Certainly to have you turned Parson will be a change— I expect I shall want it enlarging not contracting —cos I think *you* think an apology is wanting for writing abt any thing so unimportant as the mind of man!

To William Preyer 17 February [1870]

Down. | Beckenham | *Kent. S.E.*
Feb. 17

My dear Sir

I am very much obliged for your extremely kind letter & for your several presents. Although your appreciation of my work is certainly too high, yet it is very encouraging to me, especially as yesterday I read two pamphets, just published in England, in which every form of abuse is heaped on me. I am called, for instance, a "filthy dreamer".— You seem to be doing splendid work in physiology, the noblest of sciences, as I have long thought it. What you say about the differences of the blood-crystals is truly astonishing. I am also much interested by what you say of the different effect of Prussic acid on different individuals of the same species: I remember some years ago wishing in vain for information on this head. I think it arose from observing how differently in quickness (whether due to rate of respiration or to direct action of the poison I could not tell) the vapour acted on different insects. I remember Bees died instantly, but (I think,) it was a Longicorn beetle which resisted all effects for an astonishing time.—

R. W. Darwin is my Father, but I believe he was greatly aided in his paper on Optics by his Father Erasmus D.

I have really nothing of interest about myself,, but as you desire it, I will scribble whatever occurs to me.— I derived *no* advantage from the Lectures at Edinburgh, for they were infinitely dull & cured me of any taste for Geology for 3 years. D[r] Grant was not a Professor, but worked at zoology out of pure love, & his Society was a great encouragement. I used to amuse myself with examining marine animals, but I did so solely for amusement. I believe I was the first person who ever saw the earliest locomotive egg-like state of a Bryozoon: I showed it to Grant, who stated so at the meeting of the Wernerian Nat. Hist. Soc. & this little discovery was an immense encouragement.— I was disgusted at anatomy & attended only 2 or 3 Lectures & this has been ever since an irreparable loss to me.— When I went to Cambridge, I became a most enthusiastic collector of Coleoptera; but again only for amusement. If any one told me the name of a Beetle, I thought I knew all that anyone could desire; & I believe I never ever looked even at the oral organs of any insect! Yet I worked like a slave at collecting. Henslow's Society was a great charm & benefit to me, & I liked much his Lectures on Botany. All my early life I was mad for collecting, minerals, shells, plants, Bird-skins have all had their turn. Near the end of my Cambridge life Henslow persuaded me to begin Geology. I was always very fond of observing the habits of Birds, & White's Nat. Hist. of Selborne, thus had much influence on my mind. But of all books, Humboldt's Travels had by far the greatest influence— I read large parts over & over again.— I had nearly managed to get a party to go to the Canary Isl^{ds}., when the offer of joining the *Beagle* was made to me & joyfully accepted. I suppose,, however, no [one] ever started worse prepared than I was except as a mere collector. I knew nothing of anatomy, & had never read any systematic work on Zoology— I had never touched a compound microscope & had begun Geology for only about 6 months. But I took out plenty of Books & worked as hard as I could & dissected roughly all sorts of the lower marine animals. Here I felt fearfully the want of practice & Knowledge. My education in fact began on board the Beagle. I remember nothing previously which deserved to be called education except some experimental work at chemistry when a school-boy with my Brother.— No doubt collecting largely in so many branches had improved my powers of observation.—

I never wrote so much about myself in my life, & I hope it may be worth your reading, but I doubt.—

Believe me, my dear Sir | Yours sincerely | Ch. Darwin

I do not know whether you will care to see extracts from my letters printed by Prof. Henslow, but I send a copy by this Post.—[1]

[1] Extracts from ten letters written to John Stevens Henslow during the *Beagle* voyage were read to members of the Cambridge Philosophical Society on 16 November 1835.

To H. E. Darwin [March–May? 1870]

My dear Hen.

I have worked through, (& it is hard work,) half of the 2^{d} Chapter on mind, & your corrections & suggestions are *excellent.* I have adopted the greater number, & I am sure that they are very great improvements.— Some of the transpositions are most just. You have done me real good service; but by Jove how hard you must have worked & how thoroughily you have mastered my M.S. I am pleased with this Chapter now that it comes fresh to me.—

Your affectionate, admiring & obedient | Father, C. D.

All is as clear as daylight— your plan of putting corrections saves me a world of trouble, by just as much as it must have caused you.— N.B. you **can** write, I see, a perfectly clear hand, as in *all* the corrections.—

[Francis Galton undertook to test CD's hypothesis of pangenesis by carrying out blood transfusions in rabbits and seeing whether the colour of the offspring of the tranfused rabbits (silver-greys) was influenced by the colour of the donor rabbits (which had white markings). Galton failed to confirm CD's hypothesis by this means.]

From Francis Galton 12 May 1870

42 Rutland Gate SW
May 12 1870

My dear Darwin

Good rabbit news.! One of the latest litters has a white forefoot. It was born April 23rd. but as we do not disturb the young, the forefoot was not observed till to-day. the little things had huddled together shewing only their backs & heads and the foot was never suspected. The mother was injected from a grey and white and the father from a black and white.

This, recollect, is from a transfusion of only $\frac{1}{8}$th part of alien blood in Each parent; now, after many unsuccesful experiments, I have greatly improved the method of operation and am beginning on the other jugulars of my stock. Yesterday I operated on 2 who are doing well to-day & who now have $\frac{1}{3}$rd. alien blood in their veins. On Saturday I hope for still greater success. and shall go on at any waste of rabbit life, until I get at least $\frac{1}{2}$ alien blood. The experiment is not fair to Pangenesis until I do.

We are for the time relieved from anxiety about my poor dear mother, who suffered the agonies of death over & over again, but has strangely pulled through, & is now comfortable though very weak and seriously shaken

Very sincerely yours | Francis Galton

To Thomas Henry Farrer 13 [May 1870]

Down, Beckenham | Kent
13th

My dear Mr Farrer …

I am endeavouring to persuade Mr. Bruce to have inserted in Census query whether in each household the parents are cousins: I am deeply convinced that this is an important subject: if you can influence any member of government, pray do so. Some few M.P.s will take up the question.— I have given my reasons in a Chapt in 2d. Vol. of my Domestic animals.—

Pray believe me | Yours very sincerely | C. Darwin

We were both truly grieved to hear of M^rs. Farrer's most serious illness, though I hope she is now quite well again.

To Albert Günther 15 May [1870]

Down. | Bromley. | Kent. S.E..
May 15th.—

My dear Dr Günther

Sincere thanks.— Your answers are wonderfully clear & complete. I have some analogous questions on Reptiles &c. which I will send in a few days & then I think I shall cause no more trouble.— I will get the Books, you refer me to.— The case of the Solenostoma is magnificent, so exactly analogous to that of those birds in which the female is the more gay, but ten times better for me, as she is the incubator.— As I crawl on with the successive classes I am astonished to find how similar the rules are about the nuptial or "wedding dress" of all animals. The subject has begun to interest me in an extraordinary degree; but I must try not to fall into my common error of being too speculative. But a drunkard might as well say, he would drink a little & not too much! My essay, as far as fishes, batrachians & reptiles are concerned, will be in fact yours, only written by me.—

With hearty thanks | Yours very sincerely | Ch. Darwin

Remember whenever so inclined how glad we shall be to see you here.—

Thanks for proof-plates; I shall be very glad indeed to see the paper.—

[In his book, *Contributions to the theory of natural selection* (London and New York, 1870), Alfred Russel Wallace announced that he no longer believed that natural selection could account for the moral and spiritual development of humans.]

From H. W. Bates 20 May 1870

Royal Geographic Society | 15, Whitehall Place, S.W.
May 20 1870

My dear Mr Darwin

I have been having some conversation with the Editor of the "Academy" about Mr Wallace's last book & the appearance of backsliding from the Darwinian theory which it contains. Other sincere friends of the pure truth have expressed a little surprise & bewilderment at the same phenomenon. The views of friend Wallace are so plausible & suit so well wide-spread prejudices that you no doubt think with me they ought to be controverted. But who is to criticise them? No one but yourself. I do not think anyone else would have the present insight into the fallacy but yourself: to others it would require much study & labour to marshal the arguments. I said so to Mr Appleton & he begged of me to write to you in support of a request he is going to make to you to write him a short article as review of the book.

When you were last in town I spoke to you about some sentences I had written on Man, interwoven in last chapter of Mrs Somerville's book.[1] It weighs on my conscience to think that you took too much notice of what I said: for I do not really think there is much in the matter worthy of your attention

Yours sincerely | H W Bates

[1] Mary Somerville, *Physical geography*, 6th edition revised by H. W. Bates (London, 1870).

To H. W. Bates [22 May 1870]

Cambridge
Sunday

My dear Bates

I have heard from Mr Appleton, but I have written to him to say that I am very sorry that I must decline undertaking the Review, which however, I fully grant is very desirable.— I have had no practice so it w^d. take me much time; nor do I know that I shd. succeed. But my chief reason is that I really have not a grain of spare strength, & have come down here now for 2 or 3 days to try & rest. You wd. not readily believe how often & urgently I am pressed to write articles— a week ago a most urgent request from one who had much claim: shortly before another: & it is an immense relief to me to be able to say that I *never* write reviews.— So I am very sorry, but must decline, even though you recommend it—

My dear Bates | Yours very sincerely | Ch. Darwin.

P.S. If you can let me see an old Proof of your Sentences on Man, I sh^d **really very much** like to read them.—

To J. D. Hooker 25 May [1870]

Down. | Beckenham | *Kent. S.E*
May 25

My dear Hooker

I have indeed had time to grow very old since we have written to each other: never so long before. I had intended more than once writing, but I heard indirectly how hard you had been pressed by work & so refrained. I suppose thank God that your British Flora is as good as finished. Now you have rewarded my virtue by a very long & pleasant letter, telling me much that I wished to hear. I am heartily glad that Willy has come back so well & not in the least deteriorated in manners; not that I sh^{d}. have expected this from all that I have heard of him. It will be a fearful puzzle for you what to do with him now: God knows it is puzzle enough in every case whatever; & what our Boy Horace is to do, I know no more than the man in the moon: he is clever enough; but so often ailing in health, that I doubt whether he can stick to anything. About ourselves I have little news: Henrietta has just returned much strengthened by her 4 months residence at Cannes. Last Friday we all went to the Bull Hotel at Cambridge to see the Boys & for a little rest & enjoyment.— The Backs of the Colleges were simply paradisical. On Monday I saw Sedgwick who was most cordial & kind: in the morning I thought that his mind was enfeebled; in the evening he was brilliant & quite himself. His affection & kindness charmed us all. My visit to him was in one way unfortunate; for after a long sit he proposed to take me to the Museum; & I could not refuse, & in consequence he utterly prostrated me; so that we left Cambridge next morning, & I have not recovered the exhaustion yet. Is it not humiliating to be thus, killed by a man of 86, who evidently never dreamed that he was Killing me.— As he said to me "Oh I consider you as a mere baby to me".— I saw Newton[1] several times, & several nice friends of Frank.—[2] But Cambridge without dear Henslow was not itself: I tried to get to the two old Houses, but it was too far for me.—

As for myself I have been working away on Man, & as usual running to *much* greater length than I expected. I hope to go to press this autumn. In the plants line I have been continuing my comparison of crossed & self-fertilised plants, & am coming I think to some interesting results, & some curious anomalies.— Do you happen by any strange chance to have seeds of *Canna Warszewiczi* **matured in some hot country**; they w^{d}. be of great value to me— I shall, however, receive some from Italy.— …

Will you not come here some Sunday—or any time, any day w^{d} suit us, & the sooner the better: you w^{d} rejoice all our hearts. . . .

Yours affect. | C. Darwin

[1] Alfred Newton.
[2] Francis Darwin.

[James Crichton-Browne was the director of the West Riding Asylum. CD received much information from him for *Expression*, published in 1872. The book discussed in this letter was CD's copy of Guillaume Duchenne's *Mécanisme de la physionomie humaine, ou analyse électrophysiologique de l'expression des passions* (Paris, 1862).]

From James Crichton-Browne 6 June 1870

West Riding Asylum | Wakefield
6th. June 1870.

My dear Sir,

I am ashamed to write to you & infinitely distressed to contemplate all the annoyance & trouble that my negligence may have occasioned you. Enclosure was given to my man-servant to pack up. He did so & placed it in cupboard where it was again forgotten.

Today I have *myself* seen it despatched by rail carriage—prepaid. Will you kindly let me have one line to say whether it arrives in safety.

Enclosed in Duchenne (at the beginning) you will find a few crude notes on expression. I promise more, in a little time, although I fear you will scarcely trust to me after all my carelessness. Bear in mind, in extenuation of my faults that I am one of the hardest worked men in her Majestys Dominions. As a rule I *toil* daily from 8. a.m. to 11. pm. contending all the while with bad health & great anxiety.

I send you a photograph of a female patient in the Southern Counties asylum, Dumfries N.B. under the care of D^{r}. Gilchrist in whom the bristling of the hair was well seen. The woman was in a tranquil mood when the portrait was taken. When she was agitated—the ascendant emotion being horror.— the hair stood out like *wire*.

We are beginning to take large photographs here, the size of Duchennes & will I think secure some interesting observations. I shall send you some. Is there any point connected with expression that you would particularly wish to have illustrated.?

With sincere apologies & profound esteem, | I am, | Yours most faithfully | J. Crichton-Browne

From Robert Cecil, marquess of Salisbury 7 June 1870

Hatfield House, | Hatfield, | Herts.
June. 7. 1870.

Sir

It is my pleasing duty to inform you that the University of Oxford proposes to express its deep admiration for your great achievements in the field of Natural Science, by conferring upon you an Honorary Degree at the approaching Commemoration.[1]

I trust that it will not be disagreeable to you to accept this token of their esteem: & that you will be able to be present in the theatre for that purpose.

The degree will be conferred on the morning of Tuesday June the 21st. The ceremony will be over before one.

I am | Your obedient Servant | Salisbury.

[1] CD declined to accept the honorary degree on grounds of ill health (honorary degrees were normally only awarded in person). A subsequent vote at Oxford on whether to confer the degree in absentia was tied, and the degree was not awarded.

To James Crichton-Browne 8 June [1870]

Down,
June 8

My dear Sir

Duchenne arrived this morning all safe. The loss of the book was beginning to cause me trouble, but I assure you I felt more annoyment at troubling you so much and so often than at the want of the book. Considering how hard you are worked and that you have causes of anxiety, I have more reason to apologise to you, than you to me for the accidental hiding of the book and forgetfulness of your servant. I have just been reading your remarks with very great interest: you always tell me exactly the things which I am anxious to hear. I agree with all that you say, and am particularly pleased at your remarks on the pyramidal of the nose and the so-called muscle of lasciviousness. I believe it to be all fancy. In order to test Duchenne's plates I have shown the most characteristic (hiding any indication of what they were meant to express) to between 20 and 30 persons of all kinds, and have recorded their answers: when all or nearly all agree in their answer, I trust him. Now, I believe, not *one* person understood the supposed meaning of the contracted pyramidal! As for the lascivious muscle, I did not think it worth exhibiting. I have been very glad to see the photograph of the woman with bristling hair: I suppose I might, if I wished, have a wood-cut from it: she looks like a Papuan. You propose to send me a photograph of a case of "*general paralysis of the insane*", and I should be very glad to see it: I have been trying to get a London Photographer to make me one of a young baby screaming or crying badly; but I fear he will not succeed. I much want a woodcut of a baby in this state. I presume it will be hopeless, from constant movement, to get an insane person photographed, whilst crying bitterly. Should you ever have time to send me any more notes, I can assure you that they are *fully* appreciated by me. My present book has grown so large, that I am going to take the MS. to London to see how big a book it will make; and perhaps I shall print this first, and retain what I am *now* writing on expression for a separate essay, which I will print as soon as I have got the rest of my MS. printed off.[1] To return to the Photographs; if ever you get one of a person in a paroxysm of fear or horror, I should much like to see it. Have you ever noticed whether the alæ of the nostrils are then raised or distended?

With the most sincere thanks for all your assistance, I remain | My dear Sir | yours very faithfully | Ch. Darwin

Heaven only knows whether my essay will be worth the trouble which I have caused to many of my kind friends.

[1] *Descent* was published in 1871, and *Expression* in 1872.

From T. H. Huxley 22 June 1870

Geological Survey of England and Wales

My dear Darwin

I sent the books to Queen Anne St[t] this morning— Pray keep them as long as you like, as I am not using them—

I am greatly disgusted that you are coming up to London this week—as we shall be out of Town next Sunday— It is the rarest thing in the world for us to be away and you have pitched upon the one day Cannot we arrange some other day?

I wish you could have gone to Oxford, not for your sake, but for theirs—

There seems to have been a tremendous shindy in the Hebdomadal board about certain persons who were proposed; and I am told that Pusey came to London to ascertain from a trustworthy friend who were the blackest heretics out of the list proposed—and that he was glad to assent to your being doctored, when he got back—in order to keep out seven devils worse than that first!

Ever, oh Coryphæus[1] diabolicus | your faithful follower

T. H. Huxley

June 22. 1870

[1] Coryphaeus: the leader of the chorus in a Greek play.

From J. D. Hooker 10 July 1870

Royal Gardens Kew

July 10/70

Dear Darwin …

I had a talk with the D. of Argyll last night, with whom I dined, about origin of man, & found him in a "cleft stick" about Wallace, believing him to be right in the fact about man, but allowing that he must be wrong in his argument! (he had not read that paper of Wallaces)— What a clever little beggar it is!— but I cannot follow his views about man, or quite see what he would have us to believe— His chief quarrell with the "Origin" is that you do not state that the order of evolution is preordained though he believes that you would admit this.— I told him that I did not think this was any business of your's— that you did not pretend to go into the origin of life, only into it's phenomena. I could not, before his wife & children especially, go into this matter, & avow my own (& I suppose your) belief that all speculations on preordination are utterly idle in the absence of better materials than theologies & cosmogonies supply us with— that in fact the whole subject is beyond the range of our conceptions: ...

Ever yr affec | J D Hooker

To J. D. Hooker 12 July [1870]

Down. | Beckenham | *Kent. S.E.*
July 12th

My dear Hooker ...

I have always thought the D. of Argyll wonderfully clever; but as for calling him "a little beggar" my inherited, instinctive feelings wd. declare it was a sin thus to speak of a real old Duke.—

Your conclusion that all speculation about preordination is idle waste of time is the only wise one: but how difficult it is not to speculate. My theology is a simple muddle: I cannot look at the Universe as the result of blind chance, yet I can see no evidence of beneficent design, or indeed of design of any kind in the details.—

As for each variation that has ever occurred having been preordained for a special end, I can no [more] believe in it, than that the spot on which each drop of rain falls has been specially ordained.—

Spontaneous generations seems almost as great a puzzle as preordination; I cannot persuade myself that such a multiplicity of organisms can have been produced, like crystals, in Bastian's solutions of the same kind.— I am astonished that as yet I have met with no allusion to Wyman's positive statement that if the solutions are boiled for 5 hours, no organisms appear; yet, if my memory serves me, the solutions when opened to air, immediately became stocked. Against all evidence I cannot avoid suspecting that organic particles (my **gemmules** from the separate cells of the lower creatures!) will keep alive & afterwards multiply under proper conditions. What an interesting problem it is.—

Your affect | C. Darwin

To Francis Darwin 18 October [1870]

The Moat | Tunbridge
Oct 18.

My dear Frank

I enclose cheque for 115£: I expected of course to have to aid you as you were so much longer than usual at Cambridge, but the above is rather more than I expected; & what is worse I do not see how you can hold out for this quarter with only £46.5.0.— But never, for God's sake, conceal debts from me, & tell me now, whether you owe any more: tell me this & think deliberately when you acknowledge the cheque to Down, where we return on Thursday morning.

As you have not kept accounts it is of course impossible for you [to] know how you have overspent your income.— Let me urge you to make a point of conscience (& then I know it will be done, i.e. if I can persuade you that it is a duty) to keep accounts.— You cannot be sure about paying your debts if you do not, nor can you tell how to economise & spend your income to best advantage. I have never known a man who was too idle to attend to his affairs & accounts, who did not get into difficulties; & he who habitually is in money difficulties, very rarely keeps scrupulously honourable, & God forbid that this sh^{d}. ever be your fate.— If you once got into habit of attending to your money, & this implies keeping accounts, you w^{d}. feel it very little trouble. My father, who was the wisest man I ever knew, thought it the duty of every man, young & old, to keep an account of his money; & I very unwillingly obeyed him; for I was not always so bothersome an old fellow as I daresay I appear to you.—

We have been to Leith Hill & came here yesterday & have enjoyed ourselves; though mamma is not very well to day. This is a wonderfully curious & pretty place.

Your affectionate Father | Ch. Darwin

To W. D. Fox 15 November [1870]

Down. | Beckenham | *Kent. S.E.*
Nov. 15

My dear Fox

I suppose that you are now settled for the winter (which has come with a vengeance today) in the Isle of Wight.— . . . You are a good-for-nothing man, not to have said something about your health & strength. I am a good deal used up with the excessive labour to me of correcting proofs of my present book in 2 Vols; & now it will not be finished till end of year. I owe many facts to you in the larger part viz on sexual differences of Birds. I shall be delighted to send you a copy, whenever it is published, though I have sometimes had misgivings that I ought not to do so, as I fear you will disapprove of my main conclusion on the origin of man; but I can most truly say that I have written nothing without deliberate consideration & acquiring all the knowledge which I possibly could.— I will send it, as I know well that you are a charitable man & do not without good evidence believe in bad motives in others.— It is very delightful to me to hear that you, my very old friend, like my other books, & you were one of my earliest masters in Nat. History.

I wish I had got a little more strength. I feel that each job as finished must be my last.—

Farewell | Your's affectionately | Charles Darwin

Biographical register

This list includes all correspondents and most of the persons mentioned in the letters. Following the register is a list of the main biographical sources used in its compilation.

Agassiz, Alexander (1835–1910). Swiss-born zoologist, oceanographer, and mining engineer. Son of Louis Agassiz. Began administering the affairs of the Harvard Museum of Comparative Zoology in 1860; curator from 1874. (*ANB*.)

Agassiz, Elizabeth Cabot Cary (Elizabeth) (1822–1907). Educator. A founder of the educational establishment for women that later became Radcliffe College, Cambridge, Massachusetts. President of Radcliffe College, 1894–9. Married Louis Agassiz in 1850. (*ANB*.)

Agassiz, Jean Louis Rodolphe (Louis) (1807–73). Swiss-born zoologist and geologist. Emigrated to the United States in 1846. Professor of zoology and geology, Harvard University, 1847–73. Established the Museum of Comparative Zoology at Harvard in 1859. A believer in the special and separate creation of species. (*ANB*, *DSB*.)

Ainslie, Robert. Non-conformist minister and religious writer. Resident at Tromer Lodge, Down, until 1858. Minister of Christ Church, Brighton, 1860–74. Known to the Darwins as 'the beast'. (*Correspondence* vol. 7; G. E. Evans 1897.)

Anderson, Thomas (1832–70). Scottish physician and botanist. Superintendent, Calcutta botanic garden, 1861–8; conservator of forests, Bengal, 1864–6; retired because of ill health. Instituted experiments that led to the successful cultivation of *Cinchona* in India. (Lightman ed. 2004.)

Andersson, Nils Johan (1821–80). Swedish botanist. Explored the Americas on the Eugenie expedition. (Barnhart 1965.)

Appleton, Thomas Gold (1812–84). American essayist, poet, and artist. Promoted the growth and improvement of Boston; a trustee of the Athenaeum and the Public Library; founder and benefactor of the Museum of Fine Arts. (*ANB*.)

Argyll, 8th duke of. *See* Campbell, George Douglas.

Ashburton, 2d Baron. *See* Baring, William Bingham.

Babington, Charles Cardale (1808–95). Botanist, entomologist, and archaeologist. An expert on plant taxonomy. Assistant to John Stevens Henslow, whom he succeeded as professor of botany at Cambridge University in 1861. (*ODNB*.)

Baily, John. Poulterer and dealer in live birds at 113 Mount Street, Berkeley Square, London. Author of works on the management of domestic and game fowl. (Freeman 1978, *Post Office London directory* 1868.)

Baird, Spencer Fullerton (1823–87). American zoologist and scientific administrator. Assistant secretary and curator of the Smithsonian Institution, 1850; secretary, 1878–87. Particularly interested in birds and fish. (*ANB*.)

Baker, Samuel C. and Charles N. Dealers in birds and live animals, with premises at 3 Halfmoon Passage, Gracechurch Street, and 15A Beaufort Street, Chelsea, London, and at the Rue de la Faisanderie, avenue de l'Impératrice, Paris. (*Post Office London directory* 1861.)

Bakewell, Robert (1725–95). Stockbreeder and farmer at Dishley, Leicestershire. Improved breeds of sheep and cattle. (*ODNB*.)

Balfour, John Hutton (1808–84). Scottish botanist. Professor of botany, University of Edinburgh, and regius keeper of the Royal Botanic Garden, Edinburgh, 1845–79. (*ODNB*.)

Baring, Alexander Hugh, 4th Baron Ashburton (1835–89). Statesman. MP for Thetford, 1857–67. Succeeded to the peerage in 1868. (*Modern English biography*.)

Baring, William Bingham, 2d Baron Ashburton (1799–1864). Statesman. Served as a member of parliament before succeeding to the peerage in 1848. (*ODNB*.)

Bartlett, Abraham Dee (1812–97). Superintendent of the Zoological Society's gardens, Regent's Park, 1859–97. (*ODNB*.)

Bate, Charles Spence (1819–89). Dental surgeon and authority on crustacea. (*ODNB*.)

Bates, Henry Walter (1825–92). Entomologist. Went to the Amazon with Alfred Russel Wallace, 1848–9; continued to explore the area, after Wallace returned to England, until 1859. Provided the first comprehensive scientific explanation of the phenomenon subsequently known as Batesian mimicry. Published an account of his travels, *The naturalist on the River Amazons*, in 1863. (*ODNB*.)

Beaton, Donald (1802–63). Scottish gardener. An expert on hybridisation, and regular contributor to the *Gardeners' Magazine* and the *Cottage Gardener*. (Desmond 1994.)

Becker, Lydia Ernestine (1827–90). Suffragist leader and botanist. Published *Botany for novices* (1864); awarded a Horticultural Society Gold Medal, 1865. Founder and president of the Manchester Ladies' Literary Society, 1867. Secretary to the Manchester National Society for Women's Suffrage from 1867. (*ODNB*.)

Beckles, Samuel Husband (1814–90). Barrister and palaeontologist. Discovered in the Purbeck beds the oldest known mammalian fossils. (*Modern English biography*.)

Beddome, Richard Henry (1830–1911). Army officer and botanist. Conservator of forests, Madras, 1860–82. Collected plants in India and Ceylon. (Desmond 1994.)

Bell, Thomas (1792–1880). Dental surgeon and expert on reptiles, amphibians, and crustaceans. President of the Linnean Society, 1853–61. Described the reptiles from the *Beagle* voyage. A personal friend of CD's but opposed to transmutation theory. (*ODNB*.)

Bentham, George (1800–84). Botanist. Moved his botanical library and collections to the Royal Botanic Gardens, Kew, in 1854, and was provided with facilities there for his research from 1861. President of the Linnean Society of London, 1861–74. Published *Genera plantarum* (1862–83) with Joseph Dalton Hooker. (*ODNB*.)

Bleek, Wilhelm Heinrich Immanuel (Wilhelm) (1827–75). German philologist and ethnographer. Went in 1853 to Natal with Bishop Colenso to compile a Zulu grammar; appointed interpreter to George Gray, governor of Cape Colony, 1855. Curator of the South African Public Library from 1862. (*DSAB*, *NDB*.)

Blyth, Edward (1810–73). Zoologist. Curator of the museum of the Asiatic Society of Bengal, Calcutta, India, 1841–62. (*ODNB*.)

Bonnet, Charles (1720–93). Swiss naturalist and philosopher. Discovered parthenogenesis in aphids in 1746. Studied invertebrate regeneration, entomology, and plant physiology. (*DSB*.)

Boole, Mary Everest (1832–1916). Writer and educator. Studied calculus with George Boole, whom she married in 1855. Librarian, Queen's College, London, 1865–73. Popular writer on mathematics and philosophy. Originator of Boole's Sewing Cards, an aid in teaching geometry. (*ODNB*.)

Boott, Francis (1792–1863). American-born physician and botanist. Resident in England from 1816. An early believer in the therapeutic value of fresh air. Made a study of the genus *Carex* (sedges). (*ODNB*.)

Boucher de Crèvecoeur de Perthes, Jacques (Jacques Boucher de Perthes) (1788–1868). French customs official and archaeologist. Discovered controversial evidence of early humans in the Somme River valley gravels. (*DSB*.)

Bowen, Francis (1811–90). Philosopher. Alvord Professor of natural religion, moral philosophy, and civil polity, Harvard College, 1853–89. (*ANB*.)

Brace, Charles Loring (1826–90). American philanthropist and social reformer. (*ANB*.)

Brewster, Jane Kirk (b. 1827). Second wife of David Brewster, the principal and vice-chancellor of the University of Edinburgh from 1859. (*ODNB* s.v. Brewster, David.)

Bronn, Heinrich Georg (1800–62). German palaeontologist. Professor of natural science at Heidelberg University, 1833. Translated and superintended the first German editions of *Origin* (1860) and *Orchids* (1862). (*DSB*, *NDB*.)

Brooke, James (1803–68). Army officer. After helping to suppress rebels against Brunei in Sarawak, became raja of Sarawak in 1841. (*ODNB*.)

Brown-Séquard, Charles Edouard (1817–94). French physiologist. Held professorships in France, the US, and England. Conducted pioneering research in neurology and endocrinology. (*DSB*.)

Bruce, Henry Austin, Baron Aberdare (1815–95). Statesman. Home secretary, 1868–73. (*ODNB*.)

Brullé, Gaspard Auguste (1809–73). French entomologist. Professor of entomology and comparative anatomy, Dijon University, 1839–73. (*DBF*.)

Buckle, George (1820–1900). Clergyman. Vicar of Twerton, Somerset, 1852–76. Prebendary of Wells, 1868. Rector, Weston-super-Mare, 1876–88. (*Alum. Oxon.*, *The Times*, 4 January 1900, p. 8.)

Buffon, comte de. *See* Leclerc, Georges Louis.

Bunbury, Charles James Fox, 8th baronet (1809–86). Botanist. Collected plants in South America, 1833–4; in South Africa, 1838–9, in Madeira and Tenerife, 1853. (Desmond 1994, *ODNB*.)

Busk, George (1807–86). Russian-born naval surgeon and naturalist. Served on the hospital ship at Greenwich, 1832–55. Retired from medical practice in 1855 in order to devote himself to natural history. An expert on human fossil remains. (*ODNB*.)

Butler, Samuel (1774–1839). Educationalist and clergyman. Headmaster of Shrewsbury School, 1798–1836. Bishop of Lichfield, 1836–9. (*ODNB*.)

Cameron, Julia Margaret (1815–79). Photographer. Played an important role in the governor-general's social circle in India in the 1840s. In 1860, moved to Freshwater, Isle of Wight, where she rented out to friends one of the two cottages she had bought and linked together. Took up photography in 1863, and photographed many prominent literary and scientific figures. (*ODNB*.)

Campbell, George Douglas, 8th duke of Argyll (1823–1900). Scottish statesman and author of works on science, religion, and politics. A defender of the concept of design in nature. (*ODNB*.)

Candolle, Alphonse de (1806–93). Swiss botanist, lawyer, and politician. Professor of botany and director of the botanic gardens, Geneva, from 1835. Concentrated on his own research after 1850. (*DSB*.)

Candolle, Anne Casimir Pyramus (Casimir) de (1836–1918). Swiss botanist. Son of Alphonse de Candolle; assistant to and colleague of his father. Published monographs on several families of plants. (*Proceedings of the Linnean Society of London* (1918–19): 51–2.)

Carpenter, William Benjamin (1813–85). Naturalist and university administrator. Published an early positive review of *Origin*. Made microscopical studies of Foraminifera, and played a central part in the debate about the supposed discovery of

Eozoon canadense in Canadian rocks. (*ODNB.*)

Carrière, Elie Abel (1818–96). French botanist and horticulturalist. Worked at the Muséum d'Histoire Naturelle until 1869, with an interval as head of the botanic garden, Zaragossa, Spain, in the early 1860s. Editor of the *Revue Horticole* from 1862. Worked on peach trees, dimorphism, and hybridity. (Barnhart comp. 1965, *Revue Horticole* (1896): 389–97, Tort 1996.)

Carus, Julius Victor (1823–1903). German comparative anatomist. Conservator of the Museum of Comparative Anatomy, Oxford University, 1849–51. Professor extraordinarius of comparative anatomy and director of the zoological museum, University of Leipzig, 1853. Translated the third German edition of *Origin* (1867) and, subsequently, twelve other works by CD. (*DSB.*)

Castlereagh, Lord. *See* Stewart, Robert.

Cecil, Robert Arthur Talbot Gascoyne- (Robert), 3d marquess of Salisbury (1830–1903). Statesman and prime minister. Chancellor of Oxford University from 1869. (*ODNB.*)

Clarke, Benjamin (1813–90). Systematic botanist. Devised his own system of classification. (Desmond 1994.)

Claus, Carl Friedrich (1835–99). German zoologist. Professor of zoology, Marburg, 1863; Göttingen, 1870; Vienna, 1873. Carried out research on environmental influences on variability, especially in Crustacea. A strong supporter of CD in both his writing and lecturing. (*DBE*, *NDB.*)

Cleghorn, Hugh Francis Clarke (1820–95). Physician and forester. Appointed to the Madras Medical Service, Mysore, 1842. Professor of botany, Madras, from 1852, and conservator of forests, Madras, from 1856; appointed inspector-general, 1867. (Desmond 1994.)

Clowes, William & Sons. Printers to John Murray. (*ODNB.*)

Colenso, John William (1814–83). Clergyman. Bishop of Natal, South Africa, 1853–83. Published *The Pentateuch and Book of Joshua critically examined* (1862–79), in which he argued that some passages in these books had been forged several centuries after their apparent date. There was an attempt to remove him from the bishopric in 1863, but his possession of the see was confirmed by the law courts in 1866. (*ODNB.*)

Colling, Charles (1751–1836). Stockbreeder. One of the earliest and most successful improvers of shorthorn cattle. (*ODNB.*)

Colling, Robert (1749–1820). Stockbreeder. Sold to his brother, Charles Colling, the bull from which the improved stock of shorthorns was bred. (*ODNB.*)

Crawfurd, John (1783–1868). Scottish diplomat and orientalist. Held several civil and political posts in Java, India, Siam, Cochin China, Singapore, and Burma, 1803–27. President of the Ethnological Society of London from 1861. Claimed that the theory of natural selection was of no value to ethnology except insofar as it provoked enquiry. (*ODNB.*)

Crichton-Browne, James (1840–1938). Scottish physician and psychiatrist. Medical director, West Riding Asylum, Wakefield, 1866–76. (*ODNB.*)

Croll, James (1821–90). Scottish geologist. Secretary to the Geological Survey of Scotland, 1867–80. Wrote on cosmology, on oceanic circulation patterns, and on climate change and the causes of the glacial epoch. (*ODNB.*)

Crüger, Hermann (1818–64). German botanist. Government botanist and director of the botanic garden, Trinidad, from 1857. (Desmond 1994.)

Cuvier, Jean Léopold Nicolas Frédéric (Georges) (1769–1832). French comparative anatomist and palaeontologist. Professor of natural history, Collège de France, 1800–32; professor of comparative anatomy, Muséum d'Histoire Naturelle, 1802–32. (*DSB.*)

Dallas, William Sweetland (1824–90). Entomologist, author, and translator. Translated Fritz Müller, *Für Darwin* (1869); prepared the index for *Variation* and the glossary for *Origin* 6th ed. (Freeman 1978.)

Dana, James Dwight (1813–95). American geologist and zoologist. Associate editor of the *American Journal of Science and Arts* from 1846. Professor of natural history (later geology and mineralogy), Yale University, 1855–90. A man of strong religious convictions who preached that science had much to offer religion. (*ANB.*)

Darwin, Emma (1808–96). Youngest daughter of Josiah Wedgwood II. Married CD, her cousin, in 1839. (*Emma Darwin* (1915).)

Darwin, Erasmus (1731–1802). CD's grandfather. Physician, botanist, and poet. Advanced a theory of transmutation similar to that subsequently propounded by Jean Baptiste de Lamarck. (*ODNB*.)

Darwin, Erasmus Alvey (1804–81). CD's brother. Qualified in medicine but never practised. Lived in London. (Freeman 1978.)

Darwin, Francis (1848–1925). CD's son. Botanist. BA, Trinity College, Cambridge, 1870. Collaborated with CD on several botanical projects, 1874–82. Lecturer in botany, Cambridge University, 1884. (*ODNB*.)

Darwin, Francis Sacheverel (1786–1859). Son of Erasmus Darwin by his second wife, Elizabeth Collier Pole. Justice of the peace and deputy lieutenant of Derbyshire. Knighted, 1820. (*Alum. Cantab.*)

Darwin, George Howard (1845–1912). CD's son. Mathematician. BA, Trinity College, Cambridge, 1868; fellow, 1868–78. Studied law in London, 1869–72, but did not practise. Plumian Professor of astronomy and experimental philosophy, Cambridge University, 1883–1912. (*ODNB*.)

Darwin, Henrietta Emma (1843–1927). CD's daughter. Assisted CD with some of his writing. Married Richard Buckley Litchfield in 1871. (Freeman 1978.)

Darwin, Horace (1851–1928). CD's son. Civil engineer. BA, Trinity College, Cambridge, 1874. Apprenticed to an engineering firm in Kent. Founder and director of the Cambridge Scientific Instrument Company. (*ODNB*.)

Darwin, Leonard (1850–1943). CD's son. Military engineer. Attended the Royal Military Academy, Woolwich, 1868; commissioned in the Royal Engineers, 1871. Held posts at the School of Military Engineering, Chatham, and in the War Office, 1885–90. (*ODNB*.)

Darwin, Robert Waring (1766–1848). CD's father. Physician. Had a large practice in Shrewsbury and resided at The Mount. Son of Erasmus Darwin and his first wife, Mary Howard. Married Susannah, daughter of Josiah Wedgwood I, in 1796. (Freeman 1978.)

Darwin, Susan Elizabeth (1803–66). CD's sister. Lived at The Mount, Shrewsbury, the family home, until her death. (Freeman 1978.)

Darwin, William Erasmus (1839–1914). CD's eldest son.

Banker. BA, Christ's College, Cambridge, 1862. Partner in the Southampton and Hampshire Bank, Southampton, 1861. (*Alum. Cantab.*, *Correspondence.*)

Davidson, Thomas (1817–85). Artist and palaeontologist. Fellow of the Geological Society of London. Expert on fossil brachiopods. (*ODNB.*)

Davis, Jefferson (1808–89). American statesman. President of the Confederacy. (*ANB.*)

Dayman, Joseph. Commander with the Royal Navy. (*Navy list* 1860.)

Demosthenes (384–322 BCE). Athenian orator. (*Oxford classical dictionary.*)

Devay, Francis Marie Antoine (Francis) (1813–63). French medical practitioner. Professor of clinical medicine, Ecole de Médecine de Lyon, 1854. (*DBF.*)

Dohrn, Felix Anton (Anton) (1840–1909). German zoologist. Studied at various German universities, and for a time with Ernst Haeckel. Founded the Zoological Station at Naples, which opened in 1874. The station was the first marine laboratory, and served as a model for other similar institutions throughout the world. (*DSB.*)

Draper, John William (1811–82). English-born American chemist. Emigrated to the US in 1832. Professor of chemistry in New York City. At the British Association for the Advancement of Science meeting in 1860, made a lengthy speech in which he argued that societies, like organisms, followed laws of development. Later published *History of the conflict between science and religion.* (*ANB*, *Correspondence* vol. 8.)

Duberry, Amy (b. 1834/5). Dressmaker and Sunday School teacher in Down, Kent. (Census returns 1871 (Public Record Office RG/10/875); *Correspondence* vol. 16, letter from J. B. Innes, 18 June [1868]; *Correspondence* vol. 17, letter to J. B. Innes, 18 October 1869.)

Duchenne, Guillaume Benjamin Amand (1806–75). French physician. Carried out experiments on the therapeutic use of electricity. One of the founders of neurology. (*DBF.*)

Engelmann, Georg (George) (1809–84). German-born physician and botanist. Went to the US in 1832. Made fundamental contributions to the classification and taxonomy of many plant families, especially grapes, cacti, and yuccas. (*ANB*, *DAB*.)

Engleheart, Stephen Paul (1831/2–85). Surgeon in Down, Kent, 1861–70. Resident in Shelton, Norfolk, 1870–81; in Old Calabar, Nigeria, 1882–5. (*Medical directory* 1861–86.)

Evans, John (1823–1908). Archaeologist, numismatist, and paper manufacturer. In 1859, his study of chipped flints found in the Somme valley helped to establish the antiquity of humans in western Europe. Published an important paper on the fossil bird, *Archaeopteryx*, in 1865. (*ODNB*.)

Evans, Mrs. Housekeeper at Down House.

Evans-Lombe, Elizabeth (1820–98). Sister of Joseph Dalton Hooker. Married Thomas Robert Evans-Lombe in 1853. (Allan 1967.)

Falconer, Hugh (1808–65). Scottish palaeontologist and botanist. Superintendent of the Calcutta botanic garden and professor of botany, Calcutta Medical College, 1848–55. Retired owing to ill health and returned to Britain; pursued palaeontological research while travelling in southern Europe. (*ODNB*.)

Faraday, Michael (1791–1867). Natural philosopher. Professor of chemistry at the Royal Institution of Great Britain, 1833–65. Noted for his popular lectures and for his extensive researches in electrochemistry, magnetism, and electricity. (*ODNB*.)

Farrar, Frederic William (1831–1903). Anglican clergyman and headmaster. Author of the moralising school story, *Eric, or little by little* (1858). Also wrote on philology and theology. Arranged for CD to be buried in Westminster Abbey. (*ODNB*.)

Farrer, Frances (d. 1870). Daughter of William and Maitland Erskine. Married Thomas Henry Farrer in 1854. (*ODNB* s.v. Farrer, Thomas Henry.)

Farrer, Thomas Henry (1819–99). Civil servant. Secretary of the marine department, Board of Trade, 1850, rising to permanent secretary of the Board of Trade, 1867–86. (*ODNB*.)

Fergusson, James (1808–86). Scottish writer on architecture and India merchant. Travelled widely in India and published many works on its architecture. (*ODNB*.)

FitzRoy, Robert (1805–65). Naval officer, hydrographer, and meteorologist. Commander of HMS *Beagle* on her 1831–6 voyage to survey the coastline of South America; CD sailed as his companion. Governor of New Zealand, 1843–5. From 1854, worked at the meteorological department at the Board of Trade, where he pioneered a system of weather forecasting and storm warnings that was appreciated abroad but not by his employers. (*ODNB*.)

Forbes, Edward (1815–54). Zoologist, botanist, and palaeontologist. Naturalist on board HMS *Beacon*, 1841–2. Appointed professor of botany, King's College, London, 1842. Palaeontologist with the Geological Survey of Great Britain, 1844–54. (*ODNB*.)

Ford, George Henry (1809–76). South African artist. Moved to England in 1837 and became an artist at the British Museum. Friend of Albert Günther; provided illustrations for *Descent* vol. 2. (Gunther 1972.)

Fox, William Darwin (1805–80). Clergyman. CD's second cousin. A friend of CD's at Cambridge; introduced CD to entomology. Rector of Delamere, Cheshire, 1838–73. (Freeman 1978.)

Galton, Francis (1822–1911). Statistician and scientific writer. CD's cousin. Carried out various researches into heredity. Founder of the eugenics movement. (*ODNB*.)

Gärtner, Karl Friedrich von (1772–1850). German physician and botanist. Practised medicine in Calw, Germany, from 1796, but left medical practice to pursue a career in botany. Studied plant hybridisation. (*ADB*, *DSB*.)

Gaudry, Albert-Jean (Albert) (1827–1908). French palaeontologist. Carried out excavations at Pikermi, Attica, in 1855 and 1860, and published *Animaux fossiles et géologie de l'Attique* (1862–7). Taught a course in palaeontology at the Sorbonne, 1868–71; appointed professor of palaeontology at the Muséum d'Histoire Naturelle, 1872. (*DBF*, *DSB*.)

Gegenbaur, Carl (or Karl) (1826–1903). German comparative anatomist and zoologist. A supporter of CD; emphasised the importance of comparative anatomy in evolutionary reconstruction. Professor at the university of Jena, 1856–73; Heidelberg, from 1873. (*DSB*.)

Geikie, Archibald (1835–1924). Scottish geologist. Joined the Scottish branch of the Geological Survey in 1855 and became director in 1867. Director-general of the Geological Survey of Great Britain, 1882–1901. (*ODNB*.)

Geoffroy Saint-Hilaire, Etienne (1772–1844). French zoologist. Professor of zoology, Muséum d'Histoire Naturelle, 1793. Devoted much attention to embryology and teratology. (*DSB.*)

Gilchrist, James (1813–85). Physician and botanist. Medical assistant, Crichton Royal Institution, Dumfries, 1850–3; medical superintendent, Montrose Royal Asylum, 1853–7; medical superintendent, Crichton Royal Institution, Dumfries, 1857–79. (*Transactions of the Botanical Society of Edinburgh* 17 (1886–9): 2–11.)

Godwin-Austen, Robert Alfred Cloyne (1808–84). Geologist. Noted for his work on the stratigraphy of southern England. Predicted the existence of coal-bearing strata in the south-east. Held offices in the Geological Society of London. (*ODNB.*)

Grant, Robert Edmond (1793–1874). Scottish zoologist and comparative anatomist. An early supporter of the theory of the transmutation of species. Befriended CD in Edinburgh. Professor at University College, London, 1827–74. (*ODNB.*)

Gratiolet, Louis Pierre (1815–65). French anatomist and anthropologist. Held posts at the Muséum d'Histoire Naturelle and the Faculté de Sciences, Paris. (*DSB.*)

Gray, Asa (1810–88). American botanist. Fisher Professor of natural history, Harvard University, 1842–88. Wrote numerous botanical textbooks and works on North American flora. A leading supporter of CD's theory in the US, he believed that natural selection and Protestant theology were compatible. (*ANB.*)

Gray, Jane Loring (1821–1909). A member of Boston society. Married Asa Gray in 1848. (*ANB* s.v. Gray, Asa.)

Griesbach, Alexander William (b. 1806/7). Clergyman. Student at Trinity College, Cambridge, 1827–32. (*Alum. Cantab.*)

Guizot, François Pierre Guillaume (François) (1787–1874). French historian and statesman. Professor of modern history at the Sorbonne, 1812–30. Suspended from teaching, 1822–8, because of his political writing. Held a number of offices in the interior ministry, 1814–48. (*DBF.*)

Günther, Albert Charles Lewis Gotthilf (Albert) (1830–1914). German-born zoologist. Worked at the British Museum from 1857; made catalogues of the museum's specimens of amphibia, reptiles, and fish. Keeper of the zoological department, 1875–95. Edited the *Record of Zoological Literature*, 1864–9. (*ODNB.*)

Günther, Roberta (d. 1869). Natural history artist. Née M'intosh. Married Albert Günther in 1868. Died shortly after giving birth to a son. (*ODNB* s.v. Günther, Albert.)

Haeckel, Agnes (1842–1915). Daughter of Emil Huschke. Second wife of Ernst Haeckel, whom she married in 1867. (Krauße 1987.)

Haeckel, Anna (1835–64). Née Sethe. Cousin of Ernst Haeckel, whom she married in 1862. (*DSB* s.v. Haeckel, Ernst; Uschmann 1984, p. 317.)

Haeckel, Ernst Philipp August (Ernst) (1834–1919). German zoologist. Professor extraordinarius of zoology, Jena University, 1862–5; professor of zoology and director of the Zoological Institute, 1865–1909. Specialist in marine invertebrates. Leading populariser of evolutionary theory. (*DSB*.)

Harvey, William Henry (1811–66). Irish botanist. Collected plants in South Africa. Keeper of the herbarium, Trinity College, Dublin, from 1844; professor of botany, Royal Dublin Society, 1848–66; professor of botany, Trinity College, Dublin, 1856–66. Specialist in marine algae. (*ODNB*.)

Haughton, Samuel (1821–97). Irish geologist and physiologist. Professor of geology, Dublin University, 1851–81. Calculated the age of the earth at 2000 million years, though he later reduced this figure. (*ODNB*.)

Heddle, Charles William Maxwell (Charles) (1811/12–99). Senegal-born merchant. One of the richest traders in west Africa. (*ODNB*.)

Henslow, George (1835–1925). Clergyman, teacher, and botanist. Lecturer in botany at St Bartholomew's Hospital, 1866–80. Author of a number of religious books, including *Plants of the Bible* (1907), and of children's books on natural history. Younger son of John Stevens Henslow. (*Alum. Cantab.*, Desmond 1994.)

Henslow, John Stevens (1796–1861). Clergyman, botanist, and mineralogist. CD's teacher and friend. Professor of botany, Cambridge University, 1825–61. Extended and remodelled the Cambridge botanic garden. Rector of Hitcham, Suffolk, 1837–61. (*ODNB*.)

Herschel, John Frederick William, 1st baronet (1792–1871). Astronomer and mathematician. Carried out astronomical observations at the Cape of Good Hope, 1834–8. Respected as the foremost scientist of his age. (*ODNB*.)

Hildebrand, Friedrich Hermann Gustav (Friedrich) (1835–1915). German botanist. Professor of botany, Freiburg im Breisgau, 1868–1907. Worked mainly on hybridity, dimorphism, and generation. (Correns 1916, Tort 1996.)

Holland, Henry, 1st baronet (1788–1873). Physician. Related to Josiah Wedgwood I. Physician in ordinary to Prince Albert, 1840; to Queen Victoria, 1852. President of the Royal Institution of Great Britain, 1865–73. (*DNB*, *ODNB*.)

Holm, David Milne. *See* Milne, David.

Hooker, Charles Paget (1855–1933). Third child of Frances Harriet and Joseph Dalton Hooker. Became a physician. (Allan 1967.)

Hooker, Frances Harriet (1825–74). Daughter of John Stevens Henslow. Married Joseph Dalton Hooker in 1851. Assisted her husband significantly in his published work. (Allan 1967.)

Hooker, Joseph Dalton (1817–1911). Botanist. Son of William Jackson Hooker. Friend and confidant of CD. Worked chiefly on taxonomy and plant geography. Accompanied James Clark Ross on his Antarctic expedition, 1839–43, and published the botanical results of the voyage. Travelled in the Himalayas, 1847–9. Assistant director, Royal Botanic Gardens, Kew, 1855–65; director, 1865–85. (*ODNB*.)

Hooker, William Henslow (1853–1942). Eldest child of Frances Harriet and Joseph Dalton Hooker. Civil servant, India Office, 1877–1904. (Allan 1967; *India list* 1904–5.)

Hopkins, William (1793–1866). Mathematician and geologist. Tutor in mathematics at Cambridge University. President of the Geological Society of London, 1851. Specialised in quantitative studies of geological and geophysical questions. (*ODNB*.)

Horner, Leonard (1785–1864). Scottish geologist and educationalist. A promoter of science-based education at all social levels. President of the Geological Society of London, 1845–6 and 1860–1. Father-in-law of Charles Lyell. (*ODNB*.)

Humboldt, Friedrich Wilhelm Heinrich Alexander (Alexander) von (1769–1859). Prussian naturalist, geographer, and traveller. Explored northern South America, Cuba, and Mexico, and visited the United States, 1799–1804. Travelled in Siberia in 1829. (*DSB*.)

Hutton, Frederick Wollaston (1836–1905). Geologist and

army officer. Served with the Royal Welsh Fusiliers in the Crimea and India, 1855–8. Left the army in 1866, and emigrated to New Zealand. Published a favourable review of *Origin* in *Geologist* 4 (1861): 132–6, 183–8. (*DNZB*.)

Hutton, Richard Holt (1826–97). Journalist and theologian. Originally a Unitarian, his beliefs were shaken by the teachings of F. D. Maurice, and he entered the Church of England in 1862. Proprietor and joint editor of the *Spectator*, 1861–97. (*ODNB*.)

Huxley, Henrietta Anne (1825–1915). Née Heathorn. Met Thomas Henry Huxley in Sydney, Australia, in 1847, and married him in 1855. (*ODNB* s.v. Huxley, Thomas Henry.)

Huxley, Thomas Henry (1825–95). Zoologist. Assistant-surgeon on HMS *Rattlesnake*, 1846–50, during which time he investigated Hydrozoa and other marine invertebrates. Hunterian Professor, Royal College of Surgeons of England, 1862–9. Fullerian Professor of physiology, Royal Institution of Great Britain, 1855–8, 1866–9. (*ODNB*.)

Innes, John Brodie (1817–94). Clergyman. Perpetual curate of Down, 1846–68; vicar, 1868–9. Left Down in 1862 after inheriting an entailed estate at Milton Brodie, near Forres, Scotland. (*Crockford's clerical directory*, Moore 1985.)

Jamieson, Thomas Francis (1829–1913). Scottish agriculturalist and geologist. Appointed Fordyce Lecturer on agricultural research, University of Aberdeen, 1862. Carried out notable researches on Scottish Quaternary geology and geomorphology. (*Geological Magazine* 50 (1913): 332–3.)

Janet, Paul Alexandre René (Paul) (1823–99). French philosopher. Professor of logic, Lycée Louis le Grand, Paris, 1857–64. Professor of history of philosophy, Sorbonne, from 1864. Published *Le matérialisme contemporain en Allemagne: examen du système du docteur L. Büchner* (1864), in which he criticised CD's theory of natural selection. (Tort 1996.)

Jenkin, Henry Charles Fleeming (1833–85). Engineer and university teacher. Appointed professor of civil engineering, University College, London, 1866; professor of engineering, Edinburgh University, 1868. Wrote miscellaneous papers on literature, science, and political economy. (*ODNB*.)

Jenner, William, 1st baronet (1815–98). Physician. Physician to University College Hospital, London, where he held a number of professorships. Physician to Queen Victoria. (*ODNB*.)

Jenyns, Leonard (1800–93). Naturalist and clergyman. Brother-in-law of John Stevens Henslow. Vicar of Swaffham Bulbeck, Cambridgeshire, until 1853. Declined the offer to accompany Captain Robert FitzRoy on the *Beagle*'s voyage to South America, 1831. Described the *Beagle* fish specimens. Founder and first president of the Bath Natural History and Antiquarian Field Club, 1855. (*ODNB*, s.v. Blomefield, Leonard.)

Jones, Henry Bence (1814–73). Physician and chemist. Physician to St George's Hospital, 1846–62. Secretary of the Royal Institution, 1860–72. (*ODNB*.)

Jukes, Joseph Beete (1811–69). Geologist. Local director of the Geological Survey of Ireland, 1850–67; director, 1867–9. Lecturer on geology at the Royal College of Science, Dublin, from 1854. (*ODNB*.)

Kinglake, Alexander William (1809–91). Historian and travel writer. Auther of *Eothen* (1844) and *The invasion of the Crimea* (1863–87). Liberal MP for Bridgewater, 1857–69. (*ODNB*.)

Kingsley, Charles (1819–75). Author and clergyman. Appointed professor of English, Queen's College for Women, London, 1848. Regius professor of modern history, Cambridge University, 1860–9. Rector of Eversley, Hampshire, 1844–75. Chaplain to the queen, 1859–75. (*ODNB*.)

Koch, Eduard Friedrich (Eduard) (1838–97). German publisher. Took over E. Schweizerbart'sche Verlagsbuchhandlung in 1867, after which the firm published mostly scientific titles. Published a multi-volume edition of CD's works, translated by Victor Carus. (*Biographisches Jahrbuch und deutscher Nekrolog* 2 (1898): 227.)

Lamarck, Jean Baptiste Pierre Antoine de Monet (Jean Baptiste) de (1744–1829). French naturalist. Held various botanical positions at the Jardin du Roi, 1788–93. Appointed professor of zoology, Muséum d'Histoire Naturelle, 1793. Believed in spontaneous generation and the progressive development of animal types; propounded a theory of transmutation. (*DSB*.)

Langton, Charles (1801–86). Rector of Onibury, Shropshire, 1832–41. Left the Church of England in 1841. Married Emma

Darwin's sister, Charlotte Wedgwood, in 1832. After her death, married CD's sister, Emily Catherine Darwin, in 1863. (*Alum. Oxon.*, *Emma Darwin* (1915), Freeman 1978.)

Langton, Emily Catherine (Catherine) (1810–66). CD's sister. Married Charles Langton in 1863. (*Darwin pedigree.*)

Lecky, William Edward Hartpole (1838–1903). Irish historian. His *History of the rise and influence of the spirit of rationalism in Europe* (1863) set out to examine the decay of superstition in the face of reason. (*ODNB.*)

Leclerc, Georges Louis, comte de Buffon (1707–88). French naturalist, philosopher, and mathematician. Keeper, Jardin du Roi, 1739–88. His theory of transmutation was outlined in *Histoire naturelle*, published from 1749. (*DSB.*)

Lee, Robert Edward (1807–70). United States Army soldier, later Confederate Army general. Appointed commander of Virginia's armed forces, 1861. General-in-chief of all Confederate armies, 1865. Indicted for treason but never brought to trial. (*ANB.*)

Leibniz, Gottfried Wilhelm (1646–1716). German mathematician, metaphysician, and philosopher. (*ADB*, *DSB*, *NDB.*)

Lesley, J. Peter (1819–1903). American geologist. Pastor, Congregational church, Milton, Massachusetts, 1847–52. Worked on geological surveys. Professor of mining, University of Pennsylvania, and secretary and librarian of the American Philosophical Society from 1859. (*ANB.*)

Lewes, George Henry (1817–78). Writer. Author of a biography of Goethe (1855). Contributed articles on literary and philosophical subjects to numerous journals. Editor, *Fortnightly Review*, 1865–6. Published on physiology and on the nervous system in the 1860s and 1870s. Lived with Mary Ann Evans (George Eliot) from 1854. (*ODNB.*)

Lincecum, Gideon (1793–1874). American physician and naturalist. Traded with the Choctaw and Chickasaw nations in central Mississippi, recording Choctaw legends and traditions. Practised Thomsonian (herbal) medicine, 1830–48. In 1848, settled in Texas, where he wrote on natural history topics and corresponded with naturalists in the East and in Europe. Particularly known for his work on agricultural ants. (*ANB.*)

Lincoln, Abraham (1809–65). American lawyer and statesman.

Republican president of the United States of America during the American Civil War, from 1861 until his assassination in 1865. (*ANB.*)

Loring, Charles Greely (1794–1867). American lawyer and author. Father-in-law of Asa Gray. (Dupree 1959.)

Lovegrove, Henrietta (b. 1832/3). Wife of Charles Lovegrove, who was a merchant in the City of London and churchwarden of St Mary's, Down. (Census returns 1861 (Public Record Office RG9/462: 73), Freeman 1978.)

Lubbock, Ellen Frances (1834/5–79). Née Hordern. Daughter of a Lancashire Anglican minister. Married John Lubbock in 1856. (Census returns 1861 (Public Record Office RG9/462: 75), *ODNB*, s.v. Lubbock, John.)

Lubbock, Frances Mary (b. 1844/5). Daughter of the Reverend Henry Turton of Betley, Staffordshire. Henry James Lubbock's wife. (Census returns 1871 (Public Record Office RG10/875/43), *WWW* s.v. Lubbock, Henry James.)

Lubbock, Henry James (1838–1910). Banker. Second son of John William Lubbock. High sheriff for the county of London, 1897. (*WWW.*)

Lubbock, John, 4th baronet and 1st Baron Avebury (1834–1913). Banker and naturalist. A neighbour of CD's in Down, except 1861–5, when he lived at Chislehurst in Kent. Studied entomology and anthropology. Worked at the family bank from 1849; head of the bank from 1865, when he also succeeded to the baronetcy. (*ODNB.*)

Lyell, Charles, 1st baronet (1797–1875). Scottish geologist. Uniformitarian geologist whose *Principles of geology* (1830–3), *Elements of geology* (1838), and *Antiquity of man* (1863) appeared in many editions. Appointed professor of geology, King's College, London, 1831. CD's scientific mentor and friend. (*ODNB.*)

Lyell, Mary Elizabeth (1808–73). Eldest child of Leonard Horner. Married Charles Lyell in 1832. (Freeman 1978.)

McNab, James (1810–78). Scottish botanist. Collected plants in North America, 1834. Superintendent, Caledonian Horticultural Society, 1835. Curator of the Royal Botanic Garden, Edinburgh, from 1849. (Desmond 1994.)

Malthus, Thomas Robert (1766–1834). Clergyman and political economist. Quantified the relationship between growth in population and food supplies in *An essay on the principle of population* (1798). First professor of history and political economy at the East India Company College, Haileybury, 1805–34. (*ODNB*.)

Masters, Maxwell Tylden (1833–1907). Botanist and journal editor. Lecturer on botany at St George's Hospital medical school, 1855–68. GP at Peckham from 1856. Editor of the *Gardeners' Chronicle*, 1865–1907. Active in the Royal Horticultural Society; secretary of the International Horticultural Congress, 1866. (*ODNB*.)

Matthew, Patrick (1790–1874). Scottish gentleman farmer. Author of works on political and agricultural subjects. Advanced a theory of natural selection in the 1830s. (Desmond 1994.)

Miers, John (1789–1879). Botanist and civil engineer. Travelled and worked in South America, 1819–38. Author of many papers describing South American plants. A believer in the fixity of species. (*ODNB*.)

Miklucho-Maclay, Nikolai Nikolaievich (1846–88). Russian zoologist, anthropologist, and explorer. After being expelled from the University of St Petersburg for political reasons in 1864, studied at Heidelberg, Leipzig, and Jena. Worked on the comparative anatomy of various marine organisms. (*GSE*.)

Mill, John Stuart (1806–73). Philosopher and political economist. (*DSB*, *ODNB*.)

Milne, David (1805–90). Scottish advocate and geologist. Studied earthquakes and the parallel roads of Glen Roy. Founder of the Scottish Meteorological Society. (*ODNB*, s.v. Holm, David Milne.)

Mohl, Hugo von (1805–72). German biologist. Professor of botany, University of Tübingen, 1835–72. Known for his work on the microscopic anatomy of plants and for his study of the plant cell. Co-founder of *Botanische Zeitung*, 1843. (*DSB*.)

Moseley, Henry (1801–72). Mathematician and writer on mechanics. Professor of natural and experimental philosophy and astronomy, King's College, London, 1831–44. Chaplain in ordinary to Queen Victoria from 1855. FRS 1839. (*ODNB*.)

Müller, Johannes Peter (1801–58). German comparative anatomist, physiologist, and zoologist. Became professor of anatomy and physiology at Berlin University in 1833. Foreign member, Royal Society, 1840. (*ADB*, *DSB*.)

Müller, Johann Friedrich Theodor (Fritz) (1822–97). German naturalist. Emigrated to the German colony in Blumenau, Brazil, in 1852. Taught mathematics at the Lyceum in Destêrro (now Florianópolis), 1856–67. His anatomical studies on invertebrates and work on mimicry provided important support for CD's theories. (*DSB*.)

Murray, Andrew Dickson (Andrew) (1812–78). Lawyer, entomologist, and botanist. Held various offices in the Royal Horticultural Society from 1861. An expert on insects harmful to crops. (*ODNB*.)

Murray, John (1808–92). Publisher, and author of guide-books. CD's publisher from 1845. (*ODNB* s.v. Murray family, publishers.)

Nägeli, Carl Wilhelm von (1817–91). Swiss botanist. Maintained a teleological view of evolution. Studied botany under Alphonse de Candolle at Geneva. Professor of botany, University of Freiburg, 1852; Munich, 1857. (*DSB* s.v. Naegeli, Carl Wilhelm von.)

Newton, Alfred (1829–1907). Zoologist and ornithologist. Travelled extensively throughout northern Europe and North America on ornithological expeditions, 1855–65. Editor of *Ibis*, the journal of the British Ornithologists' Union, 1865–70. Professor of zoology and comparative anatomy, Cambridge University, 1866–1907. (*ODNB*.)

Newton, Isaac (1642–1727). Mathematician and natural philosopher. (*ODNB*.)

Norton, Charles Eliot (1827–1908). American editor, literary critic, and professor of art history. Apprenticed himself in the East India trade, travelling widely in India and Europe. Coedited the *North American Review*, 1863–68; and co-founded and wrote for the *Nation*. Travelled and lived in England and continental Europe, 1868–73. (*ANB*.)

Norton, Susan Ridley (1838–72). Daughter of Sara Ashburner and the American legal theorist Theodore Sedgwick. Grew up in New York and Massachusetts. Married Charles Eliot Norton in 1862. (Turner 1999.)

O'Callaghan, Patrick. BA Trinity College, Dublin, 1822; LLB & LLD 1864. Admitted *comitatis causa* to Oxford University,

1865. (*Alum. Dublin.*, *Alum. Oxon.*)

Oken, Lorenz (1779–1851). German naturalist and leading exponent of *Naturphilosophie*. Professor of natural history and rector, Zürich University, 1832–51. (*DSB*, *NDB*.)

Oliver, Daniel (1830–1916). Botanist. Assistant in the herbarium of the Royal Botanic Gardens, Kew, 1858; librarian, 1860–90; keeper, 1864–90. Professor of botany, University College, London, 1861–88. (Desmond 1994, *List of the Linnean Society of London*, 1859–91.)

Owen, Richard (1804–92). Comparative anatomist. Hunterian Professor of comparative anatomy and physiology, Royal College of Surgeons, 1836–56. Superintendent of the natural history departments, British Museum, 1856–84. Described the *Beagle* fossil mammal specimens. (*ODNB*.)

Parker, Henry (1788–1856). Physician to the Shropshire Infirmary, 1847–50. Married Marianne Darwin, CD's sister, in 1824. (*Darwin pedigree*, *Medical directory*.)

Parker, Henry (1827–92). Fine art specialist. Scholar, Oriel College, Oxford, 1846–51; fellow, 1851–85. Son of CD's sister, Marianne Parker. (*Alum. Oxon.*, *Darwin pedigree*.)

Parker, Marianne (1798–1858). CD's eldest sister. Married Henry Parker (1788–1856) in 1824. (*Darwin pedigree*.)

Pasteur, Louis (1822–95). French chemist and microbiologist. Administrator and director of scientific studies, Ecole Normale, Paris, 1857–67; director of the laboratory of physiological chemistry, 1867–88. Professor of chemistry, the Sorbonne, Paris, 1867–74. Renowned for his work on fermentation and for experiments providing evidence against the theory of spontaneous generation. (*DSB*.)

Pengelly, William (1812–94). Archaeologist and geologist. An expert on the geology of Devon; carried out excavations at Bovey Tracey, Brixham Cave, and Kent's Hole, Torquay. Founded the Devonshire Association for the Advancement of Literature, Science, and Art, 1862. (*ODNB*.)

Pictet de la Rive, François Jules (1809–72). Swiss zoologist and palaeontologist. Professor of zoology, University of Geneva, 1835. (Gilbert 1977, Sarjeant 1980–96.)

Pouchet, Félix Archimède (1800–72). French biologist and naturalist. Director of the Muséum d'Histoire Naturelle at

Rouen. Prolific author and populariser of science. Adversary of Louis Pasteur in the debate over spontaneous generation. (*DSB.*)

Powell, Henry (1839/40–72). Vicar of Downe, 1869–72. (*Alum. Cantab.*)

Prescott, William Hickling (1796–1859). American historian of Spain, Mexico, and Peru. (*ANB.*)

Prestwich, Joseph (1812–96). Geologist and businessman. Entered the family wine business in London in 1830. Professor of geology, Oxford University, 1874–88. An expert on the Tertiary geology of Europe and human prehistory. (*ODNB.*)

Preyer, Thierry William (1841–97). German physiologist and advocate of Darwinism. Born in England, where his father was a merchant. Emigrated with his family to Germany in 1855. Professor of physiology, Jena, 1869–88. (*ODNB.*)

Pusey, Edward Bouverie (1800–82). Clergyman and theologian. Regius professor of Hebrew, Oxford University, from 1828. Associated with the Oxford Movement, the Catholic revival in the Church of England. (*ODNB.*)

Quatrefages de Bréau, Jean Louis Armand de (Armand de Quatrefages) (1810–92). French zoologist and anthropologist. Professor of anthropology, Muséum d'Histoire Naturelle, 1855. (*DSB.*)

Ramsay, Andrew Crombie (1814–91). Geologist. Appointed to the Geological Survey of Great Britain, 1841; director-general, 1872–81. Lecturer on geology at the Royal School of Mines, 1851–72. (*ODNB.*)

Reade, William Winwood (1838–75). Traveller, novelist, and journalist. Travelled in West Africa, 1861–3, 1868–70, and again in 1873 as correspondent for *The Times.* Wrote novels and travel observations. (Letter from W. W. Reade, 4 June 1870 (*Calendar* no. 7216), *ODNB.*)

Robertson, John (1811/12–75). Scottish writer and editor. Reporter on the *Morning Chronicle*; editor of the *London and Westminster Review*, 1837–40. (*Modern English biography*, *Records of Lincoln's Inn.*)

Rodwell, John Medows (1808–1900). Clergyman and orientalist. Friend and contemporary of CD's at Cambridge. Rector of St Ethelburga's, Bishopsgate. Published an English version of the Koran in 1861. (*ODNB.*)

Rolleston, George (1829–81). Physician and anatomist. Physician to Radcliffe Infirmary, Oxford, and Lee's Reader in anatomy at Christ Church, Oxford, from 1857. Linacre Professor of anatomy and physiology, Oxford University, 1860–81. (*ODNB.*)

Royer, Clémence Auguste (1830–1902). French author and economist. Studied natural science and philosophy in Switzerland. In Lausanne in 1859, founded a course on logic for women. Translated *Origin* into French in 1862. (Harvey 1997.)

Sabine, Edward (1788–1883). Astronomer, geophysicist, and army officer. General secretary of the British Association for the Advancement of Science, 1839–52 and 1853–9. President of the Royal Society of London, 1861–71. (*ODNB.*)

Saint-Hilaire, Augustin François César Prouvençal (Auguste) de (1779–1853). French naturalist. Surveyed the flora and fauna of Brazil, 1816–22. Appointed professor at the faculty of sciences, Paris, 1830. (*DSB.*)

Saporta, Louis Charles Joseph Gaston (Gaston), comte de (1823–96). French palaeobotanist. Specialist on the Tertiary and Jurassic flora. Wrote extensively on the relationship between climatic change and palaeobotany. (*DSB.*)

Schmidt, Eduard Oskar (Oskar) 1823–86. German zoologist. Professor of zoology and comparative anatomy, Graz, from 1857. His 1865 inaugural lecture supporting Darwinism led to conflict with the Catholic Church in Austria and sparked a wider debate between Catholic and German nationalist circles at the university. (*DBE*, *OBL.*)

Schultze, Max Johann Sigismund (1825–74). German anatomist. Professor of anatomy and director of the anatomical institute, Bonn, from 1859. Founder of *Archiv für mikroskopische Anatomie*; editor, 1865–74. Noted for his work in microscopy, the reform of cell theory, and descriptive and taxonomic studies of rhizopods and sponges. (*DSB.*)

Sclater, Philip Lutley (1829–1913). Lawyer and ornithologist. One of the founders of *Ibis*, the journal of the British Ornithologists' Union, 1858; editor, 1858–65 and 1878–1912. Secretary of the Zoological Society of London, 1860–1903. (*DSB.*)

Scott, John (1836–80). Scottish botanist. Foreman of the propagating department at the Royal Botanic Garden, Edinburgh,

1859–64. Through CD's patronage emigrated to India in 1864; became curator of the Calcutta botanic garden in 1865. Seconded to the opium department, 1872–8. Carried out numerous botanical experiments and observations on CD's behalf. (*ODNB*.)

Scudder, Samuel Hubbard (1837–1911). American entomologist. Graduated from the Lawrence Scientific School in 1862; assistant to Louis Agassiz, 1862–4. Librarian and custodian of collections, Boston Society of Natural History, 1864–70. (*ANB*.)

Sedgwick, Adam (1785–1873). Geologist and clergyman. Woodwardian Professor of geology, Cambridge University, 1818–73. Prebendary of Norwich Cathedral, 1834–73. Made a geological tour of north Wales with CD in 1831. (*ODNB*.)

Sharpey, William (1802–80). Scottish physiologist. Professor of anatomy and physiology, University College, London, 1836–74. Secretary of the Royal Society of London, 1853–72. (*ODNB*.)

Siebold, Karl Theodor Ernst von (1804–85). German zoologist and doctor. Professor of physiology, Breslau, 1850; of physiology and comparative anatomy, Munich, 1853. Co-founder and editor of *Zeitschrift für wissenschaftliche Zoologie*, 1848. A supporter of Darwinian transmutation theory; did research on generation, especially parthenogenesis. (*DSB*.)

Silliman, Benjamin (1779–1864). American chemist, geologist, and mineralogist. Professor of chemistry and natural history, Yale University, 1802–53. Founder and first editor of the *American Journal of Science and Arts*, 1818. (*ANB*, *DSB*.)

Smith, John (1798–1888). Scottish gardener and pteridologist. Gardener at the Royal Botanic Garden, Edinburgh, 1818; at the Royal Botanic Gardens, Kew, 1822. Curator of the Royal Botanic Gardens, Kew, 1842–64. (Desmond 1994.)

Smith, John (1821–88). Scottish gardener. Gardener to the duke of Roxburgh; to the duke of Northumberland at Syon House, Middlesex, 1859–64. Curator, Royal Botanic Gardens, Kew, 1864–86. (Desmond 1994.)

Snow, George (1820/1–85). Coal-dealer, Down, Kent. Operated a weekly carrier service between Down and London. (Census returns 1861 (Public Record Office RG9/462: 72); gravestone inscription, Down churchyard; *Post Office directory of the six home counties* 1862.)

Sowerby, George Brettingham Jr (1812–84). Conchologist and illustrator. Assisted his father, George Brettingham Sowerby (1788–1854), in a business selling natural history specimens; succeeded to the business in 1854. Illustrated numerous works on shells. (*ODNB* s.v. Sowerby, George Brettingham the elder.)

Spencer, Herbert (1820–1903). Philosopher. From 1852, author of books and papers on transmutation theory, philosophy, and the social sciences. (*ODNB*.)

Spruce, Richard (1817–93). Botanist and schoolteacher. Collected plant specimens in the Pyrenees, 1845–6; in South America, 1849–67. Retired in poor health to Coneysthorpe, Yorkshire, where he worked on his plant collections. (*ODNB*.)

Stainton, Henry Tibbats (1822–92). Entomologist. Founder of the *Entomologist's Annual*, 1855, and of the *Entomologist's Weekly Intelligencer*, 1856. Secretary to the Ray Society, 1861–72. Co-founder, *Entomologist's Monthly Magazine*, 1864. An expert on microlepidoptera and Tineidae (clothes moths). (*ODNB*.)

Stebbing, Thomas Roscoe Rede (1835–1926). Clergyman and marine biologist. Fellow of Worcester College, Oxford, 1860–8. Moved to Torquay, where he worked as a tutor and schoolmaster, in 1867. Convinced by the arguments in *Origin*, he later used Darwinism as a basis from which to criticise tenets of the established church. Expert on amphipod crustaceans. (*ODNB*.)

Stewart, Robert, Viscount Castlereagh and 2d marquis of Londonderry (1769–1822). Statesman. Chief secretary of Ireland, 1798–1801. President, (East India) Board of Control, 1802–5. Secretary of state for the War and Colonial Department, 1805; secretary of state for war, 1807–9. Foreign secretary, 1812–22. Uncle of Robert FitzRoy. Committed suicide. (*ODNB*.)

Tegetmeier, William Bernhard (1816–1912). Journalist and naturalist. Pigeon and poultry editor of the *Field*, 1864–1907. Secretary of the Apiarian Society of London. (*Field*, 23 November 1912, p. 1070.)

Tennyson, Alfred, 1st Baron Tennyson (1809–92). Poet. Poet laureate from 1850. (*ODNB*.)

Thomson, William, Baron Kelvin (1824–1907). Scientist and inventor. Professor of natural philosophy, Glasgow, 1846–99.

Pioneered telegraphic systems and assisted in the laying of the first transatlantic cable. Wrote on the age and cooling of the earth. (*ODNB.*)

Timbs, John (1801–75). Author and editor. Edited the *Mirror of literature*, 1827–38; *Literary world*, 1839–40; and the *Year book of facts in science and art*, which he originated in 1839. Prolific author of popular literature. (*ODNB.*)

Tylor, Edward Burnett (1832–1917). Anthropologist. Took up anthropology after meeting Henry Christy in 1856 while travelling in the US. Interested in the origins of human culture and the laws of its development. Published *Primitive culture*, 1871. Keeper of Oxford University Museum, 1883; reader in anthropology, Oxford University, 1884. (*ODNB.*)

Victoria Adelaide Mary Louisa, Princess Royal of Great Britain and Empress of Germany (1840–1901). Eldest child of Queen Victoria and Prince Albert. Married Prince Frederick William of Prussia in 1858. (*ODNB.*)

Vogt, Carl (1817–95). German naturalist. Worked with Louis Agassiz in Switzerland on a treatise on freshwater fish. Professor of zoology, Giessen, 1846. Forced to leave the German Federation for political reasons in 1849. Professor of geology, Geneva, 1852. (*DSB*, Judel 2004.)

Wallace, Alfred Russel (1823–1913). Naturalist. Collector in the Amazon, 1848–52; in the Malay Archipelago, 1854–62. Independently formulated a theory of evolution by natural selection in 1858. Lecturer and author of works on protective coloration, mimicry, and zoogeography. (*ODNB.*)

Wallace, Violet (b. 1869). Daughter of Annie and Alfred Russel Wallace. Schoolteacher. (Raby 2001.)

Walsh, Benjamin Dann (1808–69). Entomologist, farmer, and timber merchant. Student at Trinity College, Cambridge, 1827–31. Emigrated to the United States, where he farmed in Henry County, Illinois, and then became a lumber merchant at Rock Island, Illinois, until 1857. Thereafter concentrated on entomology, making contributions to agricultural entomology. Editor of the *Practical Entomologist*, 1866–8; co-editor of the *American Entomologist* with C. V. Riley, 1868–9. Acting state entomologist, Illinois, 1867. (*Alum. Cantab.*, *ANB.*)

Wedgwood, Frances (Fanny Frank) (d. 1874). Née Mosley. Daughter of the vicar of Rolleston, Staffordshire. Married

Francis Wedgwood in 1832. (Freeman 1978.)

Wedgwood, Caroline Sarah (1800–88). CD's sister. Married Josiah Wedgwood III, her cousin, in 1837. (Freeman 1978.)

Wedgwood, Hensleigh (1803–91). Philologist. Emma Darwin's brother. An original member of the Philological Society, 1842. Published *A dictionary of English etymology* (1859–65). (*ODNB*.)

Wedgwood, Katherine Elizabeth Sophy (Sophy) (1842–1911). Daughter of Caroline Sarah Wedgwood and Josiah Wedgwood III. CD's niece. (Freeman 1978.)

Wedgwood, Lucy Caroline (1846–1919). Daughter of Caroline Sarah Wedgwood and Josiah Wedgwood III. CD's niece. (B. Wedgwood and Wedgwood 1980.)

Wedgwood, Margaret Susan (1843–1937). Daughter of Caroline Wedgwood and Josiah Wedgwood III. CD's niece. Married Arthur Charles Vaughan Williams in 1869. Mother of Ralph Vaughan Williams. (*Darwin pedigree*.)

Wedgwood, Sarah Elizabeth (Elizabeth) (1793–1880). Emma Darwin's sister. Resided at Maer Hall, Staffordshire, until 1847, then at The Ridge, Hartfield, Sussex, until 1862. Moved to London before settling in Down in 1868. (*Emma Darwin* (1915), Freeman 1978.)

Weir, Harrison William (1824–1906). Painter and illustrator. Specialised in landscape and natural history subjects. Brother of John Jenner Weir. (*ODNB*.)

Weir, John Jenner (1822–94). Naturalist and accountant. Worked in HM customs as an accountant, 1839–85. Studied entomology, especially microlepidoptera; conducted experiments on the relationships between insects and insectivorous birds. (*Science Gossip* n.s. 1 (1894): 49–50.)

Wells, William Charles (1757–1817). American-born physician. Elected assistant physician to St Thomas's Hospital, London, 1795; physician, 1800–17. Awarded the Rumford Medal of the Royal Society for his 'Essay on dew' (1814). Wrote *An account of a female of the white race of mankind, part of whose skin resembles that of a negro* (1818), which has been thought to anticipate CD's theory of natural selection. (*ODNB*.)

Westwood, John Obadiah (1805–93). Entomologist and palaeographer. Founding member of the Entomological Society of London, 1833. Hope Professor of zoology, Oxford University, 1861–93. (*ODNB*, *Transactions of the Entomological Society of*

London 1 (1833–6): xxxiv.)

White, Gilbert (1720–93). Naturalist and clergyman. Author of *The natural history and antiquities of Selborne* (1789). (*ODNB*.)

Wilberforce, Samuel (1805–73). Clergyman. Bishop of Oxford, 1845–69; Winchester, 1869–73. (*ODNB*.)

Wilkes, Charles (1798–1877). American naval officer and explorer. In 1861, sparked off the '*Trent* affair' by forcibly removing two Confederate diplomatic commissioners from a British ship. (*ANB*.)

Woodhouse, Alfred J. Dentist at 1 Hanover Square, London. The Darwin family dentist. (CD's Account book (Down House MS), *Post Office London directory*.)

Woodward, Samuel Pickworth (1821–65). Naturalist. First-class assistant in the department of geology and mineralogy, British Museum, 1848–65. Published *A manual of mollusca* (1851–6). (*ODNB*.)

Wright, Charles (1811–85). American botanical collector. Botanist to the United States North Pacific Exploring and Surveying Expedition, 1853–6. Investigated the botany of Cuba, 1856–67. (*ANB*.)

Wyman, Jeffries (1814–74). American comparative anatomist and ethnologist. Hersey Professor of anatomy, Harvard College, 1847–74. Curator, Peabody Museum of Archaeology and Ethnology, Harvard, 1866–74. (*ANB*.)

Yarrell, William (1784–1856). Zoologist. Newspaper agent and bookseller in London. Author of standard works on British birds and fishes. (*ODNB*.)

Bibliography of biographical sources

ADB: *Allgemeine deutsche Biographie.* Under the auspices of the Historical Commission of the Royal Academy of Sciences. 56 vols. Leipzig: Duncker & Humblot. 1875–1912.

Allan, Mea. 1967. *The Hookers of Kew, 1785–1911.* London: Michael Joseph.

Alum. Cantab.: *Alumni Cantabrigienses. A biographical list of all known students, graduates and holders of office at the University of Cambridge, from the earliest times to 1900.* Compiled by John Venn and J. A. Venn. 10 vols. Cambridge: Cambridge University Press. 1922–54.

Alum. Dublin.: *Alumni Dublinenses. A register of the students, graduates, professors and provosts of Trinity College in the University of Dublin (1593–1860).* New edition with supplement. Edited by George Dames Burtchaell and Thomas Ulick Sadleir. Dublin: Alex. Thom & Co. 1935.

Alum. Oxon.: *Alumni Oxonienses: the members of the University of Oxford, 1500–1886: … with a record of their degrees. Being the matriculation register of the university.* Alphabetically arranged, revised, and annotated by Joseph Foster. 8 vols. London and Oxford: Parker & Co. 1887–91.

ANB: *American national biography.* Edited by John A. Garraty and Mark C. Carnes. 24 vols. and supplement. New York and Oxford: Oxford University Press. 1999–2002.

Barnhart, John Hendley, comp. 1965. *Biographical notes upon botanists … maintained in the New York Botanical Garden Library.* 3 vols. Boston, Mass.: G. K. Hall.

Calendar: *A calendar of the correspondence of Charles Darwin, 1821–1882. With supplement.* 2d edition. Edited by Frederick Burkhardt *et al.* Cambridge: Cambridge University Press. 1994.

Correns, C. 1916. Friedrich Hildebrand. *Berichte der deutschen botanischen Gesellschaft* 34 (pt 2): 28–49.

Correspondence: *The correspondence of Charles Darwin.* Edited by Frederick Burkhardt *et al.* 15 vols to date. Cambridge: Cambridge University Press. 1985–.

Crockford's clerical directory: *The clerical directory, a biographical and statistical book of reference for facts relating to the clergy and the church. Crockford's clerical directory etc.* London: John Crockford [and others]. 1858–1900.

Darwin pedigree: *Pedigree of the family of Darwin.* Compiled by H. Farnham Burke. N.p.: privately printed. 1888. [Reprinted in facsimile in *Darwin pedigrees*, by Richard Broke Freeman. London: printed for the author. 1984.]

DBE: *Deutsche biographische Enzyklopädie.* Edited by Walter Killy *et al.* 12 vols. in 14. Munich: K. G. Saur. 1995–2000.

DBF: *Dictionnaire de biographie Française.* Under the direction of J. Balteau *et al.* 20 vols. and 2 fasciles (A–Lecompte-Boinet) to date. Paris: Librairie Letouzey & Ané. 1933–.

Desmond, Ray. 1994. *Dictionary of British and Irish botanists and horticulturists including plant collectors, flower painters and garden designers.* New edition, revised with the assistance of Christine Ellwood. London: Taylor & Francis and the Natural History Museum. Bristol, Pa.: Taylor & Francis.

DNB: *Dictionary of national biography.* Edited by Leslie Stephen and Sidney Lee. 63 vols. and 2 supplements (6 vols.). London: Smith, Elder & Co. 1885–1912. *Dictionary of national biography 1912–90.* Edited by H. W. C. Davis *et al.* 9 vols. London: Oxford University Press. 1927–96.

DNZB: *A dictionary of New Zealand biography.* Edited by G. H. Scholefield. 2 vols. Wellington, New Zealand: Department of Internal Affairs. 1940. *The dictionary of New Zealand biography.* Edited by W. H. Oliver *et al.* 5 vols. Auckland and Wellington, New Zealand: Department of Internal Affairs [and others]. 1990–2000.

DSAB: *Dictionary of South African biography.* Edited by W. J. de Kock *et al.* 4 vols. Pretoria and Cape Town: Nasionale Boekhandel Beperk [and others]. 1968–81.

DSB: *Dictionary of scientific biography.* Edited by Charles Coulston Gillispie and Frederic L. Holmes. 18 vols. including index and supplements. New York: Charles Scribner's Sons. 1970–90.

Dupree, Anderson Hunter. 1959. *Asa Gray, 1810–1888.* Cambridge, Mass.: Belknap Press of Harvard University.

Emma Darwin (1915): *Emma Darwin: a century of family letters, 1792–1896.* Edited by Henrietta Litchfield. 2 vols. London: John Murray. 1915.

Evans, George Eyre. 1897. *Vestiges of Protestant dissent: being lists of ministers … included in the National Conference of Unitarian, Liberal Christian, Free Christian, Presbyterian, and other … congregations.* Liverpool: F. & E. Gibbons.

Freeman, Richard Broke. 1978. *Charles Darwin: a companion.* Folkestone, Kent: William Dawson & Sons. Hamden, Conn.: Archon Books, Shoe String Press.

Gilbert, Pamela. 1977. *A compendium of the biographical literature on deceased entomologists.* London: British Museum (Natural History).

GSE: *Great Soviet encyclopedia.* Edited by Jean Paradise *et al.* 31 vols. New York: Macmillan. London: Collier Macmillan. 1973–83. [Translation of the 3d edition of *Bol'shaia Sovetskaia entsiklopediia*, edited by A. M. Prokhorov.]

Gunther, Albert E. 1972. The original drawings of George Henry Ford. *Journal of the Society for the Bibliography of Natural History* 6 (1971–4): 139–42.

Harvey, Joy. 1997. *'Almost a man of genius': Clémence Royer, feminism, and nineteenth-century science.* New Brunswick, N.J., and London: Rutgers University Press.

India list: *The East-India register and directory.* 1803–44. *The East-India register and army list.* 1845–60. *The Indian Army and civil service list.* 1861–76. *The India list, civil and military.* 1877–1895. *The India list and India Office list.* 1896–1917. London: Wm. H. Allen [and others].

Judel, Claus Günther. 2004. Der Liebigschüler Carl Vogt als Wissenschaftlicher, Philosoph und Politiker. *Giessener Universitätsblätter* 37: 51–6.

Krauße, Erika. 1987. *Ernst Haeckel.* 2d edition. Leipzig: B. G. Teubner.

Lightman, Bernard, ed. 2004. *Dictionary of nineteenth-century British scientists.* 4 vols. Bristol: Thoemmes Press.

List of the Linnean Society of London. London: [Linnean Society of London]. 1805–1939.

Medical directory: *The London medical directory … every physician, surgeon, and general practitioner resident in London.* London: C. Mitchell. 1845. *The London and provincial medical directory.* London: John Churchill. 1848–60. *The London & provincial medical directory, inclusive of the medical directory for Scotland, and the medical directory for Ireland, and general medical register.* London: John Churchill. 1861–9. *The medical directory . . . including the London and provincial medical directory, the medical directory for Scotland, the medical directory for Ireland.* London: J. & A. Churchill. 1870–1905.

Modern English biography: *Modern English biography, containing many thousand concise memoirs of persons who have died since the year 1850.* By Frederick Boase. 3 vols. and supplement (3 vols.). Truro, Cornwall: printed for the author. 1892–1921.

Moore, James Richard. 1985. Darwin of Down: the evolutionist as squarson-naturalist. In *The Darwinian heritage*, edited by David Kohn. Princeton, N.J.: Princeton University Press in association with Nova Pacifica.

Navy list: *The navy list.* London: John Murray; Her Majesty's Stationery Office. 1815–1900.

NDB: *Neue deutsche Biographie.* Under the auspices of the Historical Commission of the Bavarian Academy of Sciences. 22 vols. (A–Schinkel) to date. Berlin: Duncker & Humblot. 1953–.

OBL: *Österreichisches biographisches Lexikon 1815–1950.* Edited by Leo Santifaller *et al.* 11 vols. and 3 fascicles of vol. 12 (A–Slavik Ernst) to date . Vienna: Osterreichischen Akademie der Wissenschaften. 1957–.

ODNB: *Oxford dictionary of national biography: from the earliest times to the year 2000.* [Revised edition.] Edited by H. C. G. Matthew and Brian Harrison. 60 vols. and index. Oxford: Oxford University Press. 2004.

Oxford classical dictionary. Third edition. Edited by Simon Hornblower and Anthony Spawforth. Oxford: Oxford University Press. 1996.

Post Office directory of the six home counties: *Post Office directory of the six home counties, viz., Essex, Herts, Kent, Middlesex, Surrey and Sussex.* London: W. Kelly & Co. 1845–78.

Post Office London directory: *Post-Office annual directory. . . . A list of the principal merchants, traders of eminence, &c. in the cities of London and Westminster, the borough of Southwark, and parts adjacent . . . general and special information relating to the Post Office. Post Office London directory.* London: His Majesty's Postmaster-General [and others]. 1802–1967.

Raby, Peter. 2001. *Alfred Russel Wallace: a life.* London: Chatto & Windus.

Records of Lincoln's Inn: *The records of the honorable society of Lincoln's Inn: admissions.* Edited by William Paley Baildon. 2 vols. London: Lincoln's Inn. 1896.

Sarjeant, William A. S. 1980–96. *Geologists and the history of geology: an international bibliography.* 10 vols. including supplements. London: Macmillan. Malabar, Fla.: Robert E. Krieger Publishing.

Tort, Patrick. 1996. *Dictionnaire du Darwinisme et de l'evolution.* 3 vols. Paris: Presses Universitaires de France.

Turner, James. 1999. *The liberal education of Charles Eliot Norton.* Baltimore and London: The Johns Hopkins University Press.

Uschmann, Georg. 1984. *Ernst Haeckel. Biographie in Briefen.* Gütersloh: Prisma Verlag.

Wedgwood, Barbara and Wedgwood, Hensleigh. 1980. *The Wedgwood circle, 1730–1897: four generations of a family and their friends.* London: Studio Vista.

WWW: *Who was who: a companion to Who's who, containing the biographies of those who died during the period [1897–1995]*. 9 vols. and cumulated index (1897–1990). London: Adam & Charles Black. 1920–96.

Bibliographical note

The following bibliography contains a list of Darwin's works cited in the letters and notes.

Autobiography: *The autobiography of Charles Darwin 1809–1882. With original omissions restored.* Edited with appendix and notes by Nora Barlow. London: Collins. 1958.

Coral reefs: *The structure and distribution of coral reefs. Being the first part of the geology of the voyage of the* Beagle, *under the command of Capt. FitzRoy RN, during the years 1832 to 1836.* By Charles Darwin. London: Smith, Elder & Co. 1842.

Correspondence: *The correspondence of Charles Darwin.* Edited by Frederick Burkhardt *et al.* 15 vols to date. Cambridge: Cambridge University Press. 1985–.

Descent: *The descent of man, and selection in relation to sex.* By Charles Darwin. 2 vols. London: John Murray. 1871.

'Dimorphic condition in *Primula*': On the two forms, or dimorphic condition, in the species of *Primula*, and on their remarkable sexual relations. By Charles Darwin. [Read 21 November 1861.] *Journal of the Proceedings of the Linnean Society* (*Botany*) 6 (1862): 77–96. [*Collected papers* 2: 45–63.]

Expression: *The expression of the emotions in man and animals.* By Charles Darwin. London: John Murray. 1872.

'Fertilization of orchids': Notes on the fertilization of orchids. By Charles Darwin. *Annals and Magazine of Natural History* 4th ser. 4 (1869): 141–59. [*Collected papers* 2: 138–56.]

Fossil Cirripedia (1851): *A monograph on the fossil Lepadidæ, or, pedunculated cirripedes of Great Britain.* By Charles Darwin. London: Palaeontographical Society. 1851.

Fossil Cirripedia (1854): *A monograph of the fossil Balanidæ and Verrucidæ of Great Britain.* By Charles Darwin. London: Palaeontographical Society. 1854.

Living Cirripedia (1851): *A monograph of the sub-class Cirripedia, with figures of all the species. The Lepadidæ; or, pedunculated cirripedes.* By Charles Darwin. London: Ray Society. 1851.

Living Cirripedia (1854): *A monograph of the sub-class Cirripedia, with figures of all the species. The Balanidæ (or sessile cirripedes); the Verrucidæ, etc.* By Charles Darwin. London: Ray Society. 1854.

Orchids: *On the various contrivances by which British and foreign orchids are fertilised by insects, and on the good effects of intercrossing.* By Charles Darwin. London: John Murray. 1862.

Origin: *On the origin of species by means of natural selection, or the preservation of favoured races in the struggle for life.* By Charles Darwin. London: John Murray. 1859.

Origin 2d ed.: *On the origin of species by means of natural selection, or the preservation of favoured races in the struggle for life.* By Charles Darwin. London: John Murray. 1860.

Origin 3d ed.: *On the origin of species by means of natural selection, or the preservation of favoured races in the struggle for life.* With additions and corrections. By Charles Darwin. London: John Murray. 1861.

Origin 4th ed.: *On the origin of species by means of natural selection, or the preservation of favoured races in the struggle for life.* With additions and corrections. By Charles Darwin. London: John Murray. 1866.

Origin 5th ed.: *On the origin of species by means of natural selection, or the preservation of favoured races in the struggle for life.* With additions and corrections. By Charles Darwin. London: John Murray. 1869.

'Parallel roads of Glen Roy': Observations on the parallel roads of Glen Roy, and of other parts of Lochaber in Scotland, with an attempt to prove that they are of marine origin. By Charles Darwin. [Read 7 February 1839.] *Philosophical Transactions of the Royal Society of London* (1839), pt 1: 39–81. [*Collected papers* 1: 89–137.]

'Review of Bates on mimetic butterflies': [Review of 'Contributions to an insect fauna of the Amazon valley', by Henry Walter Bates.] [By Charles Darwin.] *Natural History Review* n.s. 3 (1863): 219–24. [*Collected papers* 2: 87–92.]

'Three forms of *Lythrum salicaria*': On the sexual relations of the three forms of *Lythrum salicaria.* By Charles Darwin. [Read 16 June 1864.] *Journal of the Linnean Society* (*Botany*) 8 (1865): 169–96. [*Collected papers* 2: 106–31.]

'Three sexual forms of *Catasetum tridentatum*': On the three remarkable sexual forms of *Catasetum tridentatum,* an orchid in the possession of the Linnean Society. By Charles Darwin. [Read 3

April 1862.] *Journal of the Proceedings of the Linnean Society* (*Botany*) 6 (1862): 151–7. [*Collected papers* 2: 63–70.]

'Two forms in species of *Linum*': On the existence of two forms, and on their reciprocal sexual relation, in several species of the genus *Linum*. By Charles Darwin. [Read 5 February 1863.] *Journal of the Proceedings of the Linnean Society* (*Botany*) 7 (1864): 69–83. [*Collected papers* 2: 93–105.]

Variation: *The variation of animals and plants under domestication.* By Charles Darwin. 2 vols. London: John Murray. 1868.

Zoology: *The zoology of the voyage of HMS Beagle, under the command of Captain FitzRoy RN, during the years 1832 to 1836.* Edited and superintended by Charles Darwin. 5 pts. London: Smith, Elder & Co. 1838–43.

Acknowledgments

The editors are grateful to all the institutions and individuals who have supplied copies of letters for transcription and publication for their cooperation and support. The locations of the original versions of all letters published in this volume are listed below.

American Philosophical Society, Philadelphia, Pennsylvania, USA: letters to C. Lyell, letter to L. Horner, letter to C. Kingsley, 10 June [1867], letter to J. J. Weir, letter to J. Croll, letter to T. R. R. Stebbing, letter to W. D. Fox, 15 Nov [1870].

Athenæum, 25 Apr 1863, p. 554: letter to the *Athenæum*, 18 Apr [1863].

Bath Royal Literary and Scientific Institution, Bath, England: letter to L. Jenyns.

Bayerische Staatsbibliothek, München, Germany: letter to A. Dohrn (Ana 525, Ba 697).

British Library, London, England: letters to A. R. Wallace, except for the letter to Wallace of 23 Feb 1867, which is published in J. Marchant, ed. *Alfred Russel Wallace. Letters and reminiscences* (London: 1916); letters to F. Müller (reproduced by permission of English Heritage, London, England); letter to H. E. Darwin, [before 17 Feb 1870]. By permission of the Trustees of the British Library Board.

Archives de la famille de Candolle (private collection): letter to A. de Candolle.

Christ's College Library, Christ's College, Cambridge, England: letters to W. D. Fox, 6 Feb [1867] and 21 Oct [1868].

Cambridge University Library, Cambridge, England: letters from H. W. Bates, 18 Mar 1861 (DAR 160: 61), 28 Mar 1861 (DAR 160: 62), 20 May 1870 DAR 160: 88; from L. E. Becker, 22 Dec 1866 (DAR 160: 113), 6 Feb 1867 (DAR 160: 115); from M. E. Boole, 13 Dec 1866 (DAR 160: 249), 17 Dec [1866] (DAR 160: 250); to M. E. Boole, 14 Dec 1866 (DAR 143: 121); from J. Boucher de Perthes, 23 June 1863 (DAR 160: 257); from H. G. Bronn, [before 11 Mar 1862] (DAR 160: 319); from J. M. Cameron, [before 10 July 1868] (DAR 161: 8); from A. de Candolle, 13 June 1862 (DAR 161: 10); from R. Cecil, 7 June 1870 (DAR 177: 8); from J. Crichton-Browne, 6 June 1870 (DAR 161: 311); to Crichton-Browne, 8 June [1870] (DAR 143: 332); from J. Croll, 4 Feb 1869 (DAR 161: 263); from W. S. Dallas, 8 Jan 1868 (DAR 162: 9); from Emma Darwin, [June 1861] (DAR 210.8: 35); to F. Darwin, 18 Oct [1870] (DAR 211: 6); to G. H. Darwin, 24 Jan [1868] (DAR 210.1: 3), [9 Dec 1868] (DAR 210.1: 6); to H. E. Darwin, 26 July [1867] (DAR 185: 57), [Mar–May? 1870] (DAR 185: 58); from H. E. Darwin, [17 Feb 1870] (DAR 245: 33b); to T. Davidson, 30 Apr 1861 (DAR 143: 373); from H. Falconer, 3 Jan [1863] (DAR 164: 10), (to W. Sharpey), 25 Oct 1864 (DAR 144: 475), 3 Nov 186[4] (DAR 164: 19); to F. W. Farrar, 5 Mar 1867 (DAR 144: 40–1); from F. Galton, 12 May 1870 (DAR 105: A17–18); from A. Gray, 23 Jan 1860 (DAR 98: B22–5),

22–30 Mar 1863 (DAR 165: 131), 15 and 17 May 1865 (DAR 165: 147), 24 July 1865 (DAR 165: 148); from A. Günther, 23 Sept 1869 (DAR 165: 243); from E. Haeckel, [before 6 Feb 1868] (DAR 166: 46); from J. D. Hooker, 2 July 1860 (DAR 100: 141–2), [31 Jan – 8 Feb 1862] (DAR 101: 14, DAR 111: A93), 26 Nov 1862 (DAR 101: 77–8, 61–2), 19 May 1864 (DAR 101: 220–1), 2 Dec 1864 (DAR 101: 260–1), 1 Jan 1865 (DAR 102: 1–3), 20 May 1868 (DAR 102: 210–13), 10 July 1870 (DAR 103: 53–4); to Hooker, 4 Feb [1861] (DAR 115: 87), 23 [Apr 1861] (DAR 115: 98), 23 [Apr 1861] (DAR 115: 91), 18 [May 1861] (DAR 115: 100), [after 26] Nov [1862] (DAR 115: 172), 15 and 22 May [1863] (DAR 115: 193), [27 Jan 1864] (DAR 115: 218), 4 May [1865] (DAR 115: 268), 4 Apr [1866] (DAR 115: 282), 3 and 4 Aug [1866] (DAR 115: 295), 8 Aug [1866] (DAR 115: 297), 24 Dec [1866] (DAR 115: 309), 8 Feb [1867] (DAR 94: 10–13), 17 Mar [1867] (DAR 94: 13a–e), 10 Feb [1868] (DAR 94: 50–1), 23 Feb [1868] (DAR 94: 52–4), 21 May [1868] (DAR 94: 62–4), 17 [June 1868] (DAR 94: 72–3), 14 July 1868 (DAR 94: 76–7), 28 July [1868] (DAR 94: 80–2), 23 Aug 1868 (DAR 94: 85–8), 13 Jan 1869 (DAR 94: 110–11), 22 June [1869] (DAR 94: 134–6), 25 May [1870] (DAR 94: 169–72), 12 July [1870] (DAR 94: 179–80); from H. A. Huxley, 1 Jan 1865 (DAR 166: 284); from T. H. Huxley, 20 Jan 1862 (DAR 166: 291), 1 June 1865 (DAR 166: 308), 16 July 1865 (DAR 166: 309), 22 June 1870 (DAR 166: 322); to H. B. Jones, 3 Jan [1866] (DAR 249: 86 (photocopy; by permission of the British Library); from C. Kingsley, 31 Jan 1862 (DAR 169: 29), 8 Nov 1867 (DAR 169: 37); to Kingsley, 10 June [1867] (DAR 96: 28–9), 6 Nov [1867] (DAR 270; by permission of the Charles Darwin Trust (Quentin Keynes Bequest), London, England); to a local landowner, [1866?] (DAR 96: 27r); to J. Lubbock, 5 Sept [1862] (DAR 263: 54) (by permission of English Heritage, London, England); from J. Murray, 28 Jan [1867] (DAR 171: 344); to A. Newton, 29 Mar [1864] (CUL Alfred Newton papers 177; by permission of Balfour & Newton Libraries, Department of Zoology, University of Cambridge); from W. W. Reade, 17 Jan 1869 (DAR 83: 165–6); to J. M. Rodwell, 15 Oct [1860] DAR 185: 149; to G. de Saporta, 24 Sept 1868 (DAR 147: 419); to J. Scott, 12 Nov [1862] (DAR 93: B7–B10); from Scott, 6 Jan 1863 (DAR 177: 81), 28 Mar 1864 (DAR 177: 103), 29 July [1864] (DAR 177: 112); from A. Sedgwick, 11 Oct 1868 (DAR 177: 129); from T. R. R. Stebbing, 5 Mar 1869 (DAR 177: 248); from W. B. Tegetmeier, 18 Feb 1863 (DAR 178: 57); from A. R. Wallace, 2 Jan 1864 (DAR 106: B8–11), 29 May [1864] (DAR 106: B14–19), 18 Sept 1865 (DAR 106: B25–26), 4 Feb 1866 (DAR 106: B31–32), 2 July 1866 (DAR 106: B33–B38), 24 Feb [1867] (DAR 82: A19–A22), 20 Jan 1869 (DAR 106: B73–74), 30 Jan 1869 (DAR 106: B75–76); from B. D. Walsh, 29 Apr – 19 May 1864 (DAR 181: 9), 7 Nov 1864 (DAR 181: 10); from H. W. Weir, 23 Mar 1869 (DAR 86: C10).

Dittrick Medical History Center, Cleveland, Ohio, USA: letter to H. W. Bates, 26 Mar [1861], letter to C. Kingsley, 6 Feb [1862], letter to H. W. Bates, 20 Nov [1862], letter to J. B. Innes. In a private collection at the same location is the letter to W. Preyer.

Field Museum of Natural History, Chicago, Illinois, USA: letters to B. D. Walsh.

Gardeners' Chronicle and Agricultural Gazette, 21 Apr 1860, pp. 362–3: letter to the *Gardeners' Chronicle*.

Gray Herbarium of Harvard University, Cambridge, Massachusetts, USA: letters to A. Gray.

Ernst-Haeckel-Haus, Friedrich-Schiller-Universität Jena, Germany: letters to E. Haeckel.

Houghton Library, Harvard University, Cambridge, Massachusetts, USA: letter to L. Agassiz (bMS Am 1419 (277)).

Imperial College London, London, England: letters to T. H. Huxley, and letter from Huxley to J. D. Hooker, 3 Dec 1864.

Kinnordy, Scotland (private collection): letters from L. Jenyns and T. F. Jamieson.

Linnean Society of London, Piccadilly, London, England: letter to T. H. Farrer, letter from G. Lincecum.

K. M. Lyell ed., *Life, letters and journals of Sir Charles Lyell, Bart.* (London: 1881): letters from C. Lyell.

McGill University Libraries, Rare Books and Special Collections Division, Montréal, Québec, Canada: letter to E. Blyth.

Maggs Bros Ltd (dealer), London, England: letter to H. W. Bates, [22 May 1870].

National Library of Scotland, Edinburgh, Scotland (including the John Murray Archive): letter to T. F. Jamieson (MS.5406, ff.167–8), letter from Emma Darwin to P. Matthew (Acc.10963), letters to J. Murray. By permission of the Trustees.

Archives of The New York Botanical Garden, Charles Finney Cox Collection, Bronx, New York, USA: letters to W. B. Tegetmeier.

Archives of the Royal Botanic Gardens, Kew, Richmond, Surrey, England: letter to G. Bentham.

Royal Society, London, England: letter to J. F. W. Herschel.

Mrs Romney Sedgwick (private collection): letter to A. Sedgwick.

Shrewsbury School, Shrewsbury, England: letters to A. Günther.

Staatsbibliothek zu Berlin–Preußischer Kulturbesitz, Berlin, Germany: letter to J. V. Carus (Deutsche SB Darmst. 1918.214, Carus 21).

Smithsonian Institution Libraries, Dibner Library of Science and Technology, Washington, DC, USA: letter to W. D. Fox, 24 Aug [1866].

Transactions of the Hawick Archæological Society (1908): 67–8: letter to J. Scott, 21 May [1864].

University College London, London, England: letter to F. Galton.

Wedgwood Museum Trust, Barlaston, Stoke-on-Trent, Staffordshire, England: letter from E. C. Langton (currently on deposit at Keele University Library, Keele, Staffordshire).

Manuscripts and Archives, Yale University Library, New Haven, Connecticut, USA: letter to J. D. Dana, and letter from him (Dana family papers).

Index